Excel

BASIC SKILLS

MATHS

YEAR 1

AGES 6–7

MENTAL MATHS STRATEGIES

PASCAL PRESS

Alan Parker

Reprinted 2007, 2008 (twice), 2009, 2010

Updated in 2013 for the Australian Curriculum
Reprinted 2014, 2015, 2016, 2019, 2020, 2021, 2022

Updated in 2023 for the NSW Curriculum and Australian Curriculum Version 9.0 changes

Reprinted 2024, 2025

Dedicated to my brother

ISBN 978 1 74125 184 5

Pascal Press
PO Box 250
Glebe NSW 2037
www.pascalpress.com.au

Publisher: Vivienne Joannou
Project editors: May McCool, Emma Driver and Mark Dixon
Edited by May McCool, Jeremy Billington, Valerie McCool and Mark Dixon
Answers checked by Valerie McCool
Indexed by Jeremy Billington
Original page design by Jelly Design
Typesetting by Typecellars Pty Ltd and Precision Typesetting (Barbara Nilsson)
Cover by DiZign Pty Ltd
Cover photos by photos.com
Printed by Vivar Printing/Green Giant Press

CONTENTS

Help Section

Set A

Addition

Addition is joining together of two or more groups of objects (or numbers). We use the sign + to show addition.

Hint: Other words for addition include add and plus.

e.g. and

9 altogether

e.g. plus

4 plus 4 = 8

e.g. plus

6 + 1 = 7

e.g. 6 plus 3 (answer is 9)

Addition strategies

You can add numbers in your head using strategies (or little tricks) to help you.
Let's look at some of them:

Jump strategy

e.g. 7 + 2

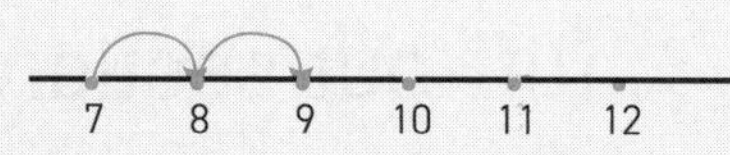

We say 7 + 1 + 1
= 9

Count on strategy

We can also use our fingers to count on by saying 7,
one finger 8,
(another finger) 9.

Doubling

e.g. 4 + 4

We can say double 4 = 8.

Near doubling

e.g. 3 + 4

We can say double 3 + 1
= 6 + 1 = 7

Related facts

When we know one addition fact, we also know another.

e.g. 3 + 4 = 7 so 4 + 3 = 7

We should know addition facts to 10. Here is a table of number facts to 10. You can use this table to help you with addition and subtraction.

+	1	2	3	4	5	6	7	8	9
1	2	3	4	5	6	7	8	9	10
2	3	4	5	6	7	8	9	10	
3	4	5	6	7	8	9	10		
4	5	6	7	8	9	10			
5	6	7	8	9	10				
6	7	8	9	10					
7	8	9	10						
8	9	10							
9	10								

Subtraction

Subtraction is taking away from a group of objects (or the difference between two numbers). We use the sign **–** to show subtraction.

Hint: Other words for subtraction include minus, take away, difference and subtract.

e.g. 

5 left

e.g. 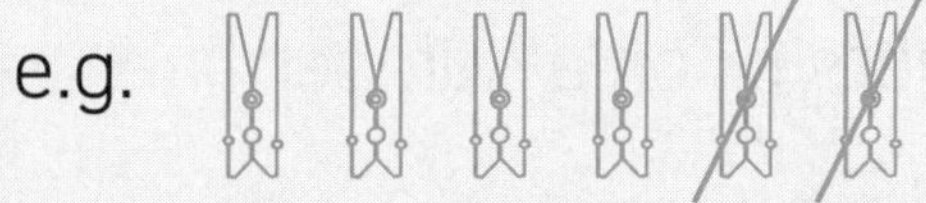

6 take away **2** = **4**

e.g.

8 minus **5** = **3**

e.g.

9 – **3** = **6**

e.g. 6 minus 5 (answer is 1)

Subtraction strategies

You can subtract numbers in your head using strategies (or little tricks) to help you.
Let's look at some of them:

Jump strategy

e.g. 9 – 3

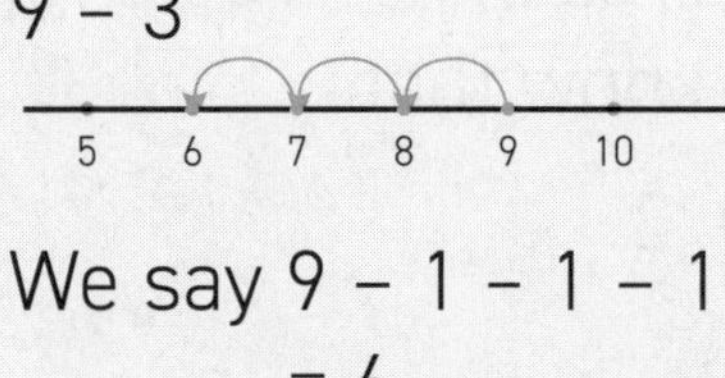

We say 9 – 1 – 1 – 1
= 6

Count back strategy

e.g. 7 – 3

We can also use our fingers to count back from 7 to 3.
We say 7
(one finger) 6,
(another finger) 5,
(another finger) 4.
We counted three fingers so that we know 7 – 3 = 4

Count on strategy

e.g. 9 – 5

We use our fingers to count from 5 to 9. We say
5 (one finger) 6,
(another finger) 7,
(another finger) 8,
(another finger) 9.
We counted a total of four fingers so we know 9 – 5 = 4

Related facts

When we know one subtraction fact, we also know another.

e.g. 7 – 4 = 3 so 7 – 3 = 4

Comparison strategy

e.g. 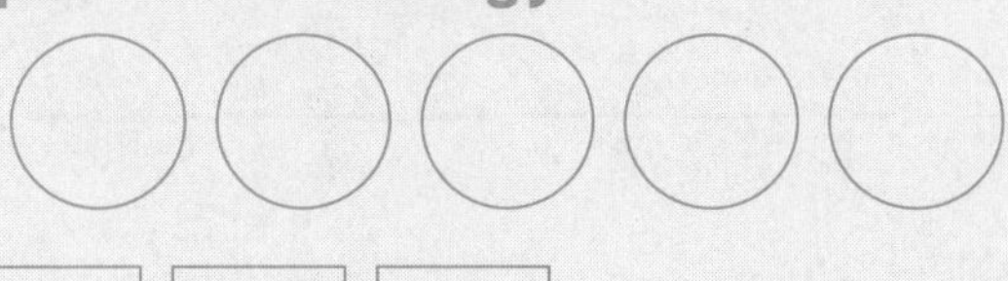

We can see that there are two more circles than rectangles so we know that

5 – 3 = 2

Multiplication

Multiplication is repeated addition of equal groups or rows of objects (or numbers). We use the sign × to show multiplication.

Hint: Other words for multiplication include times, lots of, and groups of.

e.g. 

2 rows of 4 = 8

or 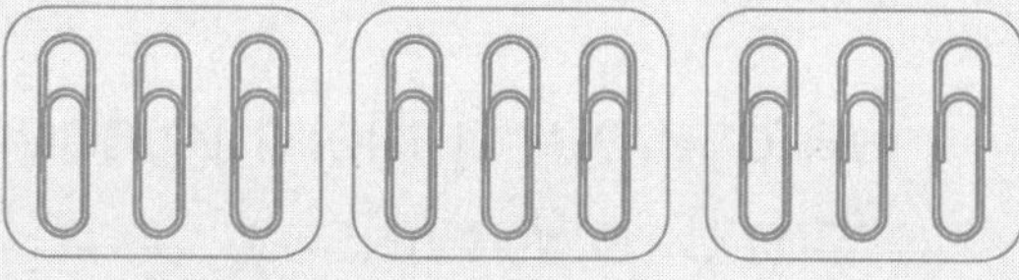

3 groups of 3 = 9

e.g.  **3** **×** **2**

(answer is 6)

Linking multiplication to addition

We can use addition to solve multiplication problems.

e.g. 2 × 4 is the same as
4 × 4 = 8

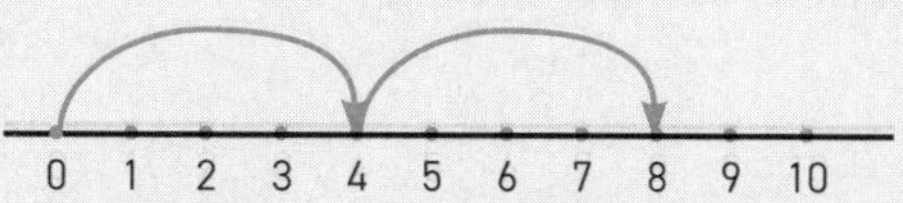

Here is a tables chart to help you with multiplication.

×	1	2	3	4	5	6	7	8	9	10
1	1	2	3	4	5	6	7	8	9	10
2	2	4	6	8	10	12	14	16	18	20
5	5	10	15	20	25	30	35	40	45	50
10	10	20	30	40	50	60	70	80	90	100

Multiplication strategies

You can multiply in your head using strategies (or little tricks) to help you.
Let's look at some of them:

Skip counting

Skip counting allows us to count on our fingers or on a number line.

e.g. 2, 4, 6

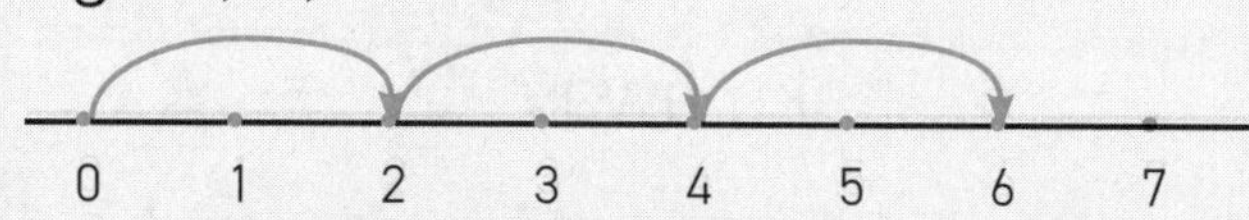

or 3 × 2 is 6.

Related facts

If you know one multiplication fact, you also know another.

e.g. 

10 dots

2 groups of 5 = 10
so 5 groups of 2 = 10

Division

Division is sharing or grouping objects into fair shares or equal groups. It is also repeated subtraction of the same number. We use the sign ÷ to show division.

Hint: Other words for division include share, how many groups of, and divided.

e.g. Share 8 between 2.
(answer is 4)

Sharing

e.g.

This is a fair share because both shares are equal.

Hint: You can share objects by using the 'one for you, one for you, and one for me' method of sharing.

e.g. Share 10 objects between two people.

Each share is 5.

Grouping

e.g. How many groups of 2 are there in 10?

There are five groups of 2.

Hint: We place objects into groups of 2 and then count the number of groups.

Division strategies

You can divide numbers in your head using strategies (or little tricks) to help you.

Let's look at some of them:

Related facts

If you know one table fact, you also know another.

e.g.

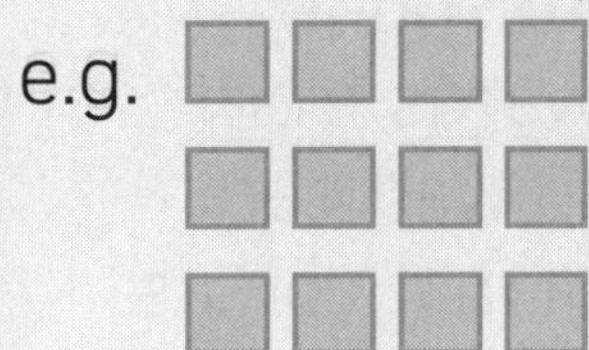

12 has 3 rows of 4
and 12 has 4 columns of 3

Set B

Numerals

Numerals are symbols used to show any given number of objects. We use the numerals 0 to 9 to make any number that we may want to show.

e.g. The numeral for

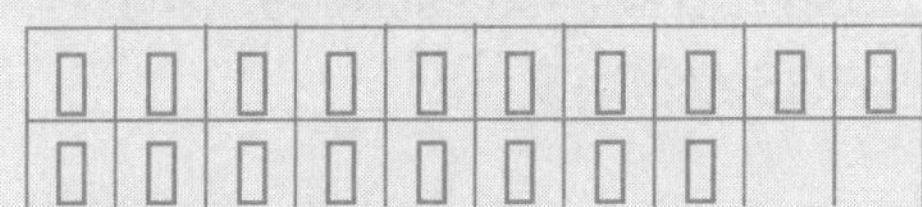

is 18 (eighteen).

e.g. The numeral for

is 30 (thirty).

e.g. The numeral for

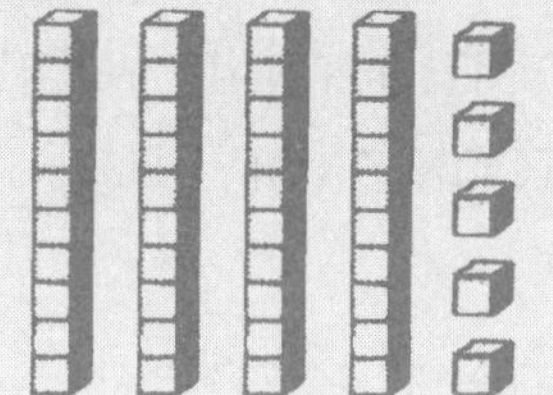

is 45 (forty-five).

Counting

We should be able to count forwards and backwards from any starting point so we need to know the number that comes before and after any number.

e.g. number before 59 is 58 and number after 59 is 60

e.g.

before		after
	30	

(answer: 29 and 31)

Numerals in words

Sometimes we need to write a numeral in words.

e.g. twenty-six

Here we write the numeral just the way it sounds when we read it as a number, 26.

e.g. Numeral for thirty-eight
(answer is 38)

Place value

We can use a numeral expander to show place value and to represent a numeral.

e.g.

Here, we can see that there are 3 tens and 7 ones in the numeral. We can write the numeral as 37.

e.g. Numeral for

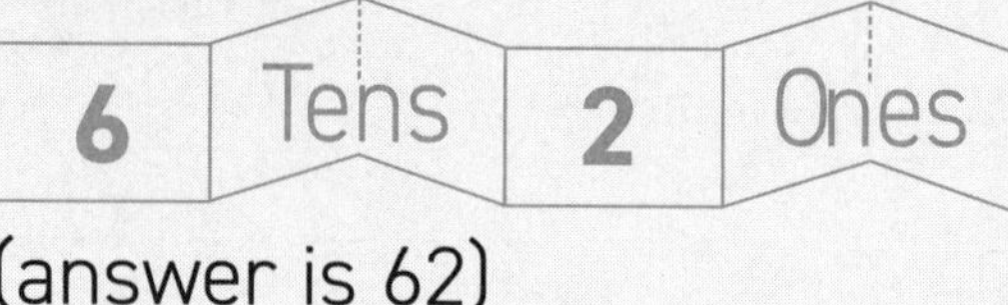

(answer is 62)

Fraction

A fraction is any part of a whole or group. We can write the common fraction as $\frac{a}{b}$ where:

a (numerator) is the number of equal fraction parts

b (denominator) is the number of equal parts that the whole has been divided into.

e.g. 1 out of 2 equal parts is a half $(\frac{1}{2})$

e.g. 1 out of 4 equal parts is a quarter $(\frac{1}{4})$

e.g. Fraction coloured

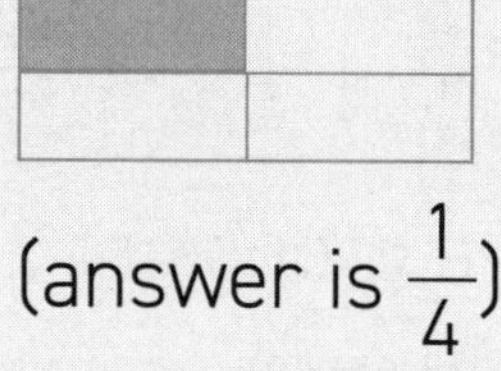

(answer is $\frac{1}{4}$)

Number pattern

A number pattern is a regular pattern made by counting or by using operations.

e.g. **2** **4** **6** **8**
counting by twos

18 **19** **20** **21**
counting forwards by ones

33 **32** **31** **30**
counting backwards by ones

We have to look closely at each number to determine the pattern of numbers.

e.g. 5 10 15 ☐

(answer is 20)

Money

We need to be able to recognise all of the coins and notes that we use in Australia. The coins and notes have different sizes, colours and patterns to help us tell them apart.

e.g. A 50 cent coin is large, silver and not round.
The 1 and 2 dollar coins are gold and round.

We could use strategies (or little tricks) to help us count money and to find the right change. Here are some of them:

Count largest to smallest

e.g.

We touch (or move) the coins and say 20, 30, 35. That's 35 cents.

Shopkeeper's counting

e.g. Change from $1 if 75c is spent.
We give out the coins as we count 75 and 5 (5c coin) is 80, and 20 (20c coin) is one dollar. The change is 25c.

e.g. Change from $1 if 55c is spent. (answer is 45c)

Ordinal number

We use ordinal numbers to order things from first to last. We use digits and abbreviations to write ordinal numbers.

e.g. 1st (first), 2nd (second), 3rd (third) etc.

e.g. Ordinal number for twelfth? (answer is 12th)

Odd number

An odd number is any number that cannot be exactly divided by 2.

e.g. 1, 3, 5, 7, 9, etc.

15

We can see that 15 is an odd number because we do not have equal pairs.

e.g. Is 21 an odd number?
(answer is yes)

Even number

An even number is any number that can be exactly divided by 2.

e.g. 2, 4, 6, 8, 10, etc.

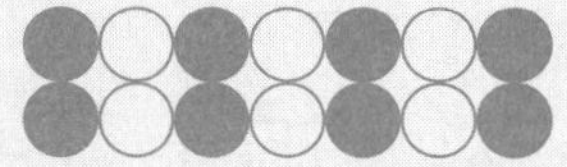

14

We can see that 14 is an even number because we have equal pairs.

e.g. Is 22 an even number?
(answer is yes)

Hundred square

You can use the hundred square in many different ways.

1	2	3	4	5	6	7	8	9	10
11	12	13	14	15	16	17	18	19	20
21	22	23	24	25	26	27	28	29	30
31	32	33	34	35	36	37	38	39	40
41	42	43	44	45	46	47	48	49	50
51	52	53	54	55	56	57	58	59	60
61	62	63	64	65	66	67	68	69	70
71	72	73	74	75	76	77	78	79	80
81	82	83	84	85	86	87	88	89	90
91	92	93	94	95	96	97	98	99	100

Jump strategy

5 more than 24 is 29

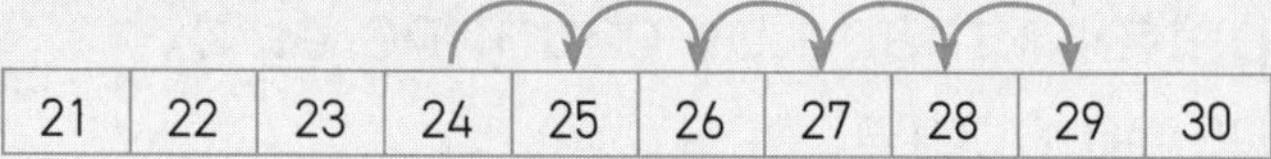

21	22	23	24	25	26	27	28	29	30

4 less than 30 is 26

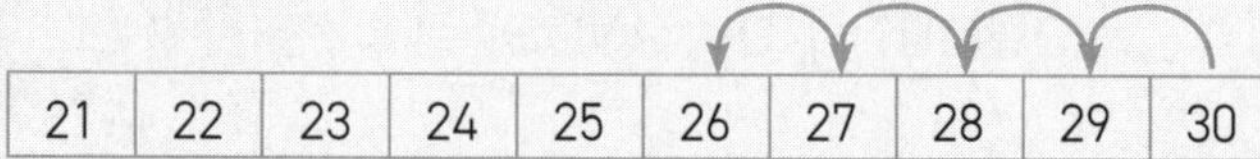

21	22	23	24	25	26	27	28	29	30

Number patterns

Even numbers

11	12	13	14	15	16	17	18	19	20
21	22	23	24	25	26	27	28	29	30
31	32	33	34	35	36	37	38	39	40

Odd numbers

11	12	13	14	15	16	17	18	19	20
21	22	23	24	25	26	27	28	29	30
31	32	33	34	35	36	37	38	39	40

Number before or after

31	32	33	34	35	36	37	38	39	40
41	42	43	44	45	46	47	48	49	50
51	52	53	54	55	56	57	58	59	60

Count by tens

31	32	33	34	35	36	37	38	39	40
41	42	43	44	45	46	47	48	49	50
51	52	53	54	55	56	57	58	59	60

Multiplication patterns

1	2	3	4	5	6	7	8	9	10
11	12	13	14	15	16	17	18	19	20
21	22	23	24	25	26	27	28	29	30
31	32	33	34	35	36	37	38	39	40

Division as sharing

1	2	3	4	5	6	7	8

Set C

Straight line

A straight line is one that takes the shortest path between two points.

e.g.

straight line

Some shapes have only straight lines.

e.g.

Curved line

A curved line is one that does not take the shortest path between two points.

e.g.

curved line

Some shapes only have a curved line.

e.g.

Sides

Shapes have different numbers of sides.

e.g. A triangle has 3 sides.

A rectangle has 4 sides.

e.g. How many sides?

(answer is 4)

Triangle

A triangle is a plane shape with three sides and three corners.

e.g.

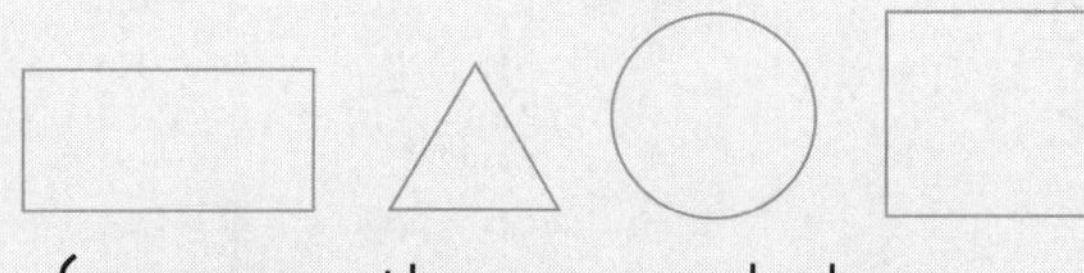

e.g. Colour the triangle.

(answer: the second shape is coloured)

Quadrilateral

A quadrilateral is a plane shape with four sides and four corners.

e.g.

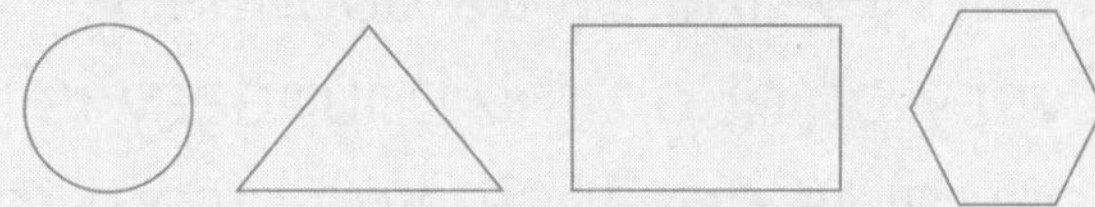

e.g. Colour the quadrilateral.

(answer: the third shape is coloured)

Rectangle

A rectangle is a quadrilateral with opposite sides equal in length and four square corners.

e.g.

e.g. Colour the rectangle.

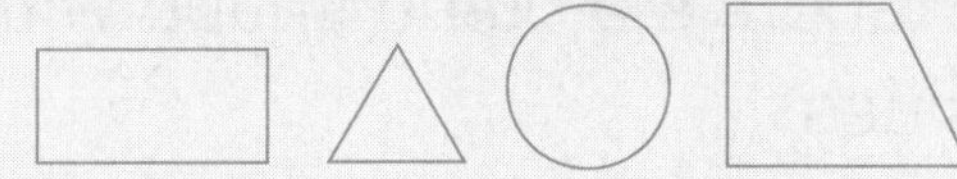

(answer: the first shape is coloured)

Square

A square is a quadrilateral with four equal sides and four square corners. A square is a special kind of rectangle.

e.g.

e.g. Colour the square.

(answer: the fourth shape is coloured)

Circle

A circle is a plane shape that has a curved line as its boundary. Every point on the boundary is the same distance away from the centre of the circle.

e.g.

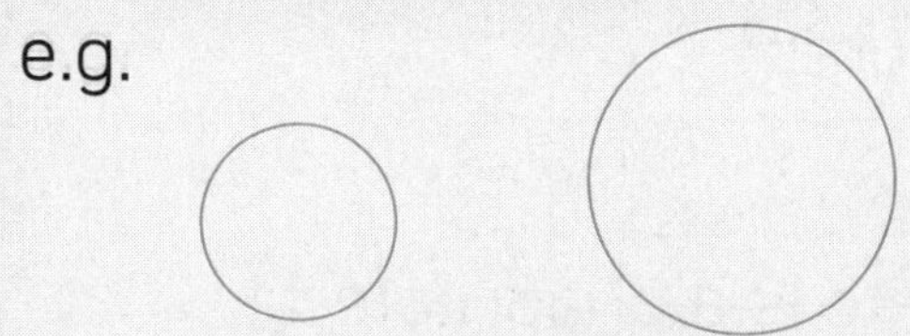

e.g. Colour the circle.

(answer: the third shape is coloured)

Hexagon

A hexagon is a plane shape with six sides.

e.g.

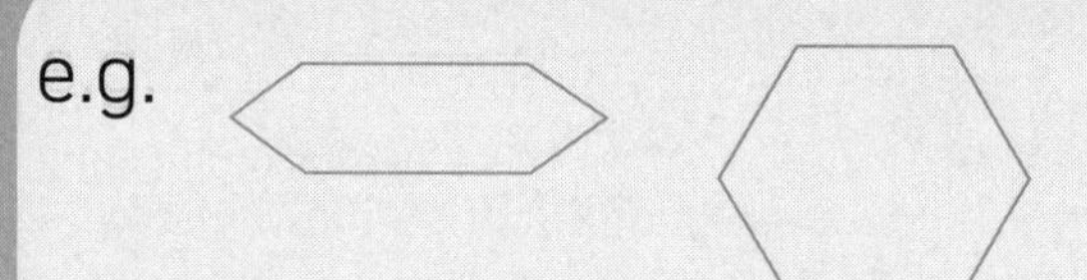

e.g. Colour the hexagon.

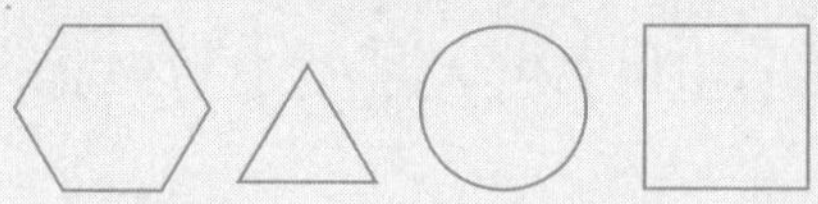

(answer: the first shape is coloured)

Octagon

An octagon is a plane shape with eight straight sides and eight corners.

e.g.

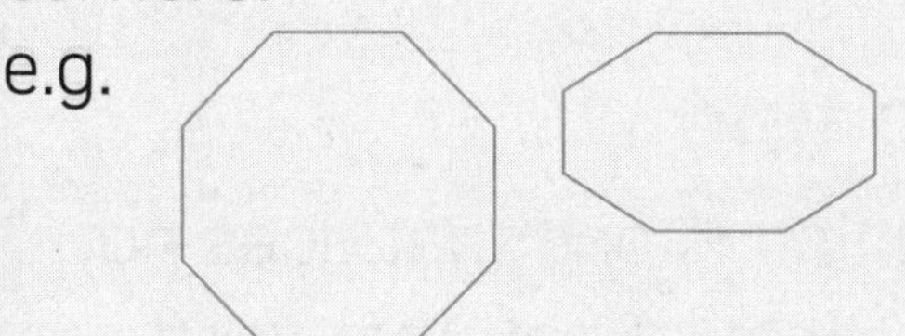

e.g. Colour the octagon.

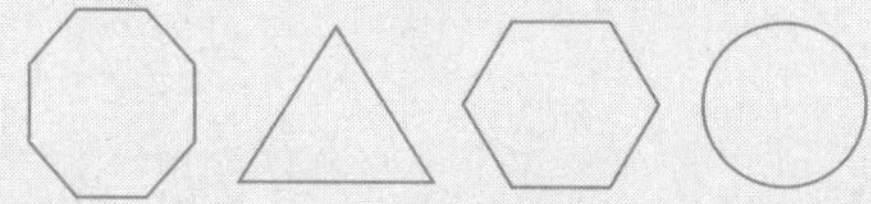

(answer: the first shape is coloured)

Diamond (rhombus)

A diamond (or rhombus) is a plane shape with four equal sides but without a square corner.
It looks like a flattened square.

e.g.

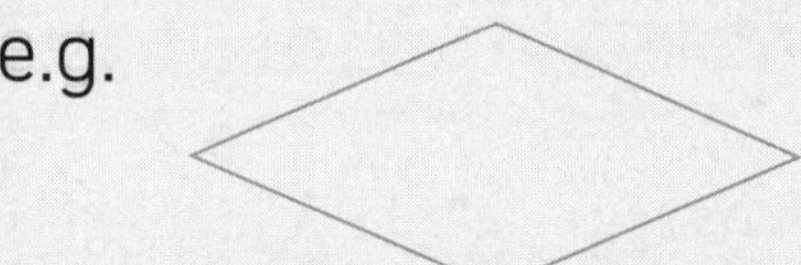

e.g. Colour the diamond.

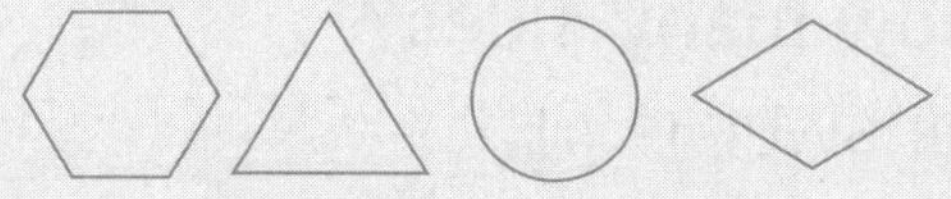

(answer: the fourth shape is coloured)

Oval

An oval is a plane shape that has a curved line at its boundary. It looks like a flattened circle.

e.g. 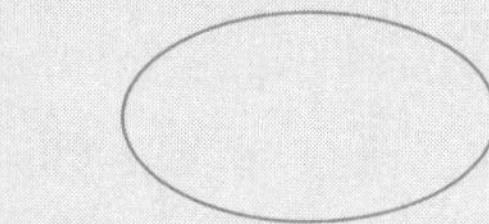

e.g. Colour the oval.

(answer: the third shape is coloured)

Solid shapes

A solid shape is a three-dimensional object. It cannot be drawn so that all parts can be seen at the same time.

Surface

A surface is an outside layer of a solid shape. It can be flat or curved.

e.g.

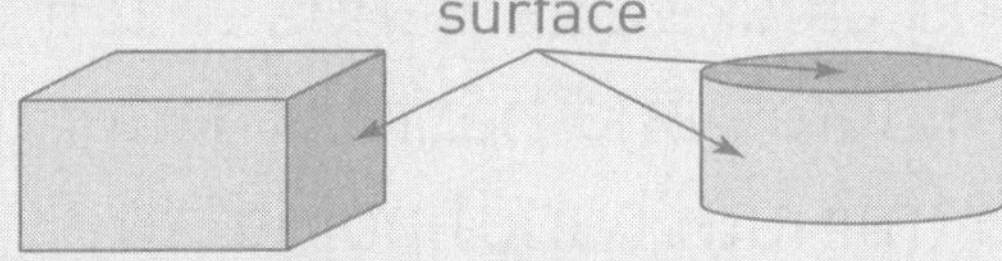

Face

A face is a flat surface.

e.g.

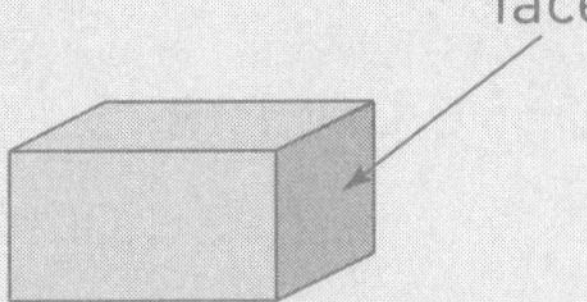

Cone

A cone is a solid shape with one flat surface and one curved surface.

e.g. 

Cylinder

A cylinder is a solid shape that has two flat surfaces and one curved surface.

e.g. 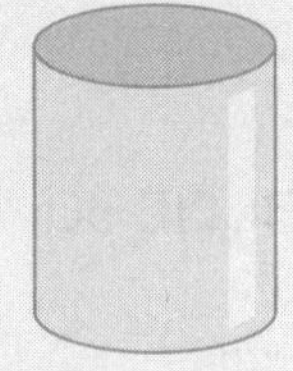 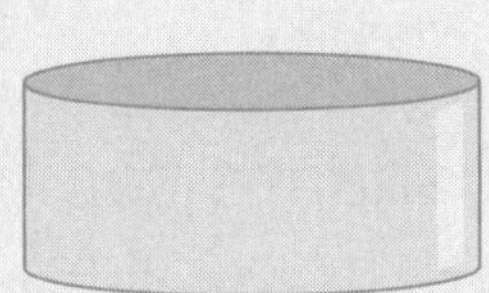

Sphere

A sphere is a solid shape that is rounded on all sides. It has only one surface and that is curved.

e.g. 

Prism

A prism is a solid shape that has only flat surfaces. It has two ends that are the same shape and all other surfaces are rectangles.

e.g. 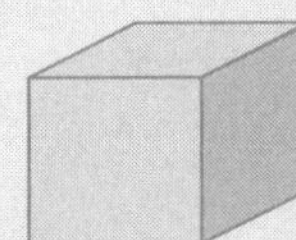 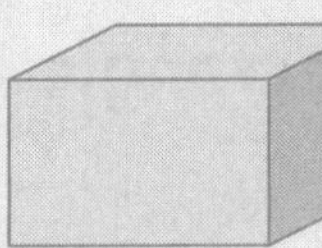 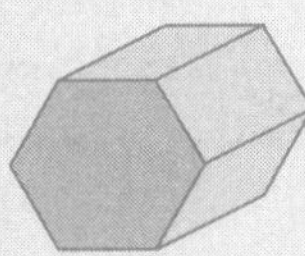

Cube

A cube is a special prism that has six square faces.

e.g. 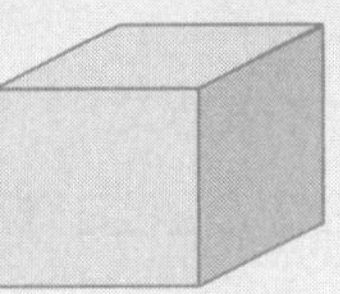

Pyramid

A pyramid is a solid shape that has only flat surfaces. It has a base and all other sides are triangles.

e.g.

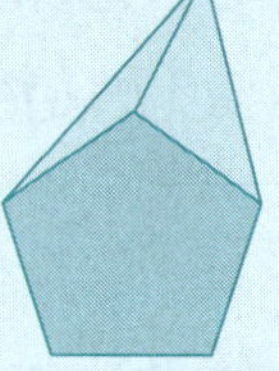

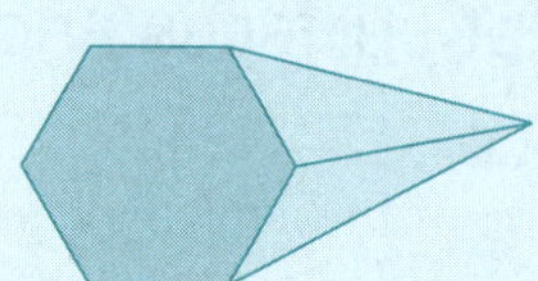

Solid shapes can be described informally by different attributes:

Shape

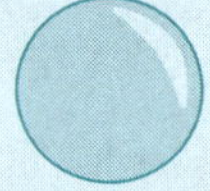

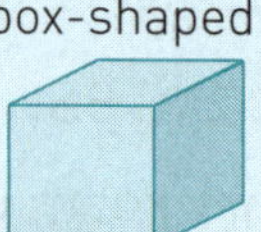

Roll, slide or stack

Position

We should know and use position words like left, right, near, behind, beside, under, etc.

e.g. Is the rectangle on the left?

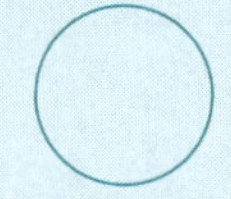

(answer is yes)

Patterns

We should be able to make, complete and describe patterns.

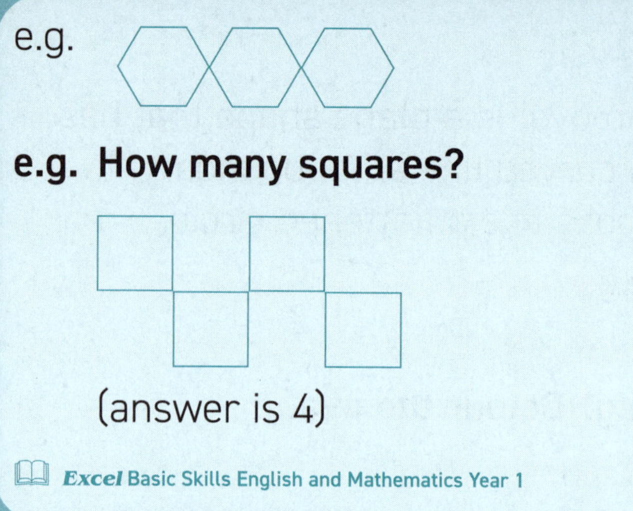

e.g.

e.g. How many squares?

(answer is 4)

Excel Basic Skills English and Mathematics Year 1

Set D

Time

We should be able to read clocks on the hour and the half-hour.

We use several different types of clocks to tell the time. Here are the most common ones found in the home and at school.

Analog clock

Analog clocks use hands to tell the minutes and hours. The long hand (minute hand) points to the minutes and the short hand (hour hand) points to the hours.

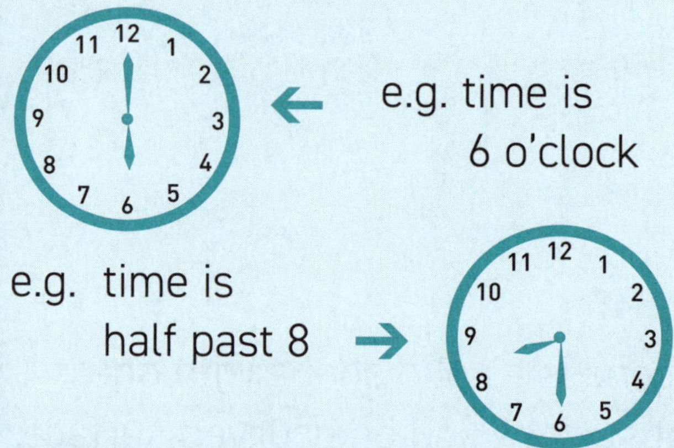

e.g. time is 6 o'clock

e.g. time is half past 8

Digital clock

Digital clocks use numerals to tell hours and minutes.

e.g. time is
10 o'clock →

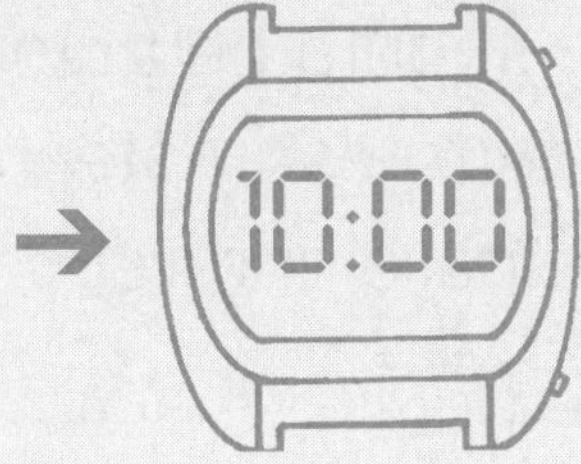

Days of the week

We should know the names of the days of the week and the order in which they occur. They are:

Sunday
Monday
Tuesday
Wednesday
Thursday
Friday
Saturday

Months of the year

We should know the names of the months and the order in which they occur. They are:

January, February,
March, April,
May, June,
July, August,
September, October,
November, December

Seasons

We should know the seasons of the year and the months that make each season in Australia.

They are:

Summer
December, January, February

Autumn
March, April, May

Winter
June, July, August

Spring
September, October, November

Days in each month

Here is a rhyme to help us remember how many days in each month.

Thirty days has September,
April, June and November.
All the rest have thirty-one
Except February alone
Which has twenty-eight days clear
And twenty-nine each leap year.

Calendar

We should be able to read a calendar to find the month, day and date.

e.g.

December

M	T	W	T	F	S	S
		1	2	3	4	5
6	7	8	9	10	11	12
13	14	15	16	17	18	19
20	21	22	23	24	25	26
27	28	29	30	31		

e.g. Date for first Saturday.
(answer is 4th)

Mass

Mass measures the amount of matter in an object but weight is commonly used in place of mass.

We should be able to use informal units to estimate, measure, compare and record the masses of two or more objects.

e.g.

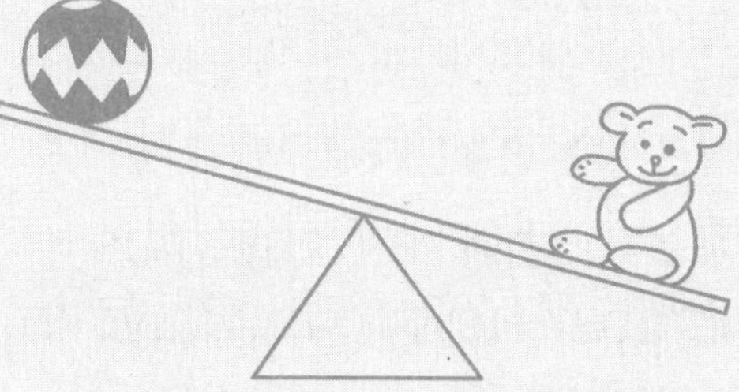

Here, we can see that the teddy is heavier than the ball.

e.g.

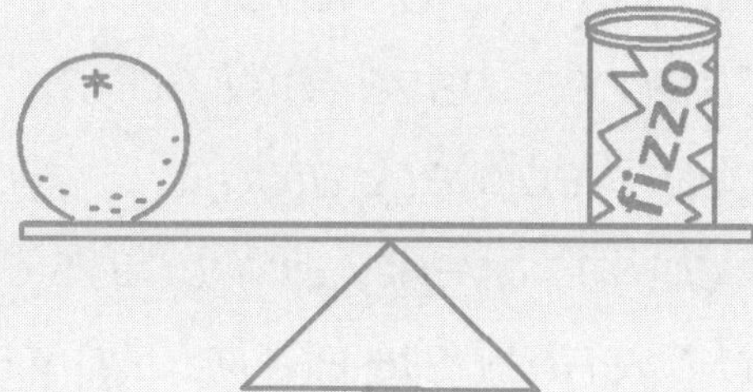

Here, we can see that the see-saw is level so the two objects have the same mass.

e.g. Circle the object that is heavier.

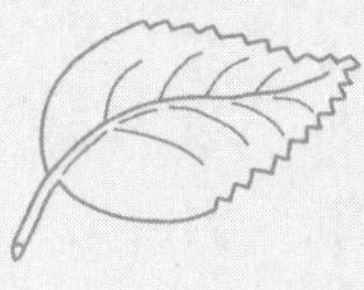

(answer: the book will be circled)

Capacity

Capacity is the amount of space inside a container.

We should be able to use informal units to estimate, measure, compare and record capacity.

e.g.

We can estimate that the jug is a larger container than the cup.

e.g. Circle the larger container.

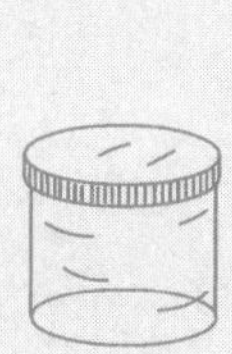

(answer is jug)

Volume

Volume is the space taken up by a solid object. We often use cubes to measure volume.

e.g.

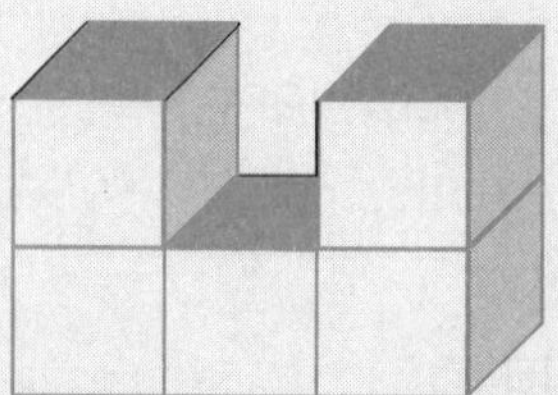

Here, we can see that the volume is 5 blocks.

e.g. How many blocks in this model?

(answer is 3 blocks)

Area

Area is the amount of space covered or enclosed by a shape.

We should be able to use informal units to estimate, measure, compare and record areas.

e.g.  

Here, we can see that the first area is larger than the second.

We can use different shapes to measure an area.

e.g. 

Here, the area is 8 rectangles.

e.g. Record the area.

(answer is 6 squares)

Length

When we talk about length, we are talking about how long something is, or the distance between two points.

We should be able to use metres and centimetres to estimate, measure, compare and record lengths.

Unit of length

metre (m)

e.g. 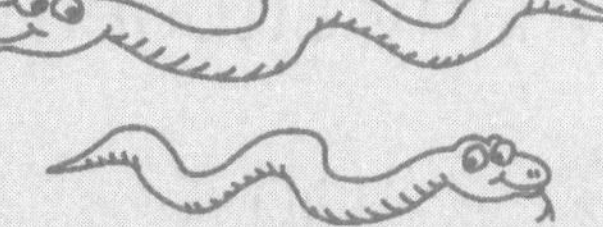

We can see that the top snake is longer than the bottom snake.

e.g. 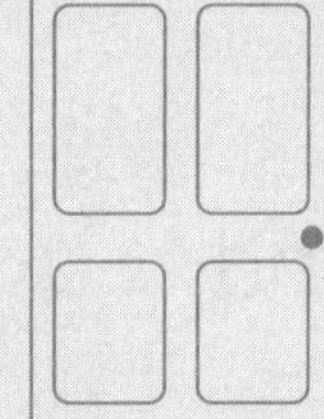 

We can estimate that a door would be longer than a finger.

e.g. 

Length is 3 pencils.

e.g. Is a door taller than 1 metre?
(answer is yes)

Picture graph

A picture graph uses separate objects to represent data. The picture graph may be drawn horizontally or vertically.

e.g. Pets kept in the classroom

How many frogs?
(answer is 3)

Unit 1

A

1 How many altogether?

a and

☐ altogether

b  and 

☐ altogether

c  and 

☐ altogether

d 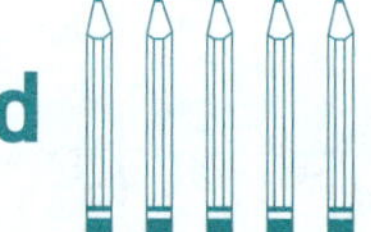 and 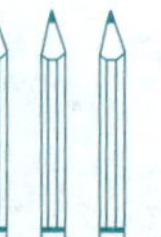

☐ altogether

e  and 

☐ altogether

B

15 is a numeral.

1 Write the numeral for:

a

★	★	★	★	★	★	★	★	★	★
★	★	★	★	★	★				

☐

b

★	★	★	★	★	★	★	★	★	★
★	★	★	★	★	★	★			

☐

c

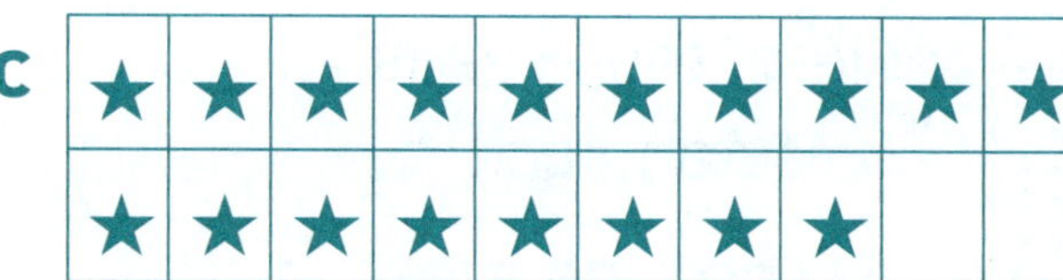

☐

d

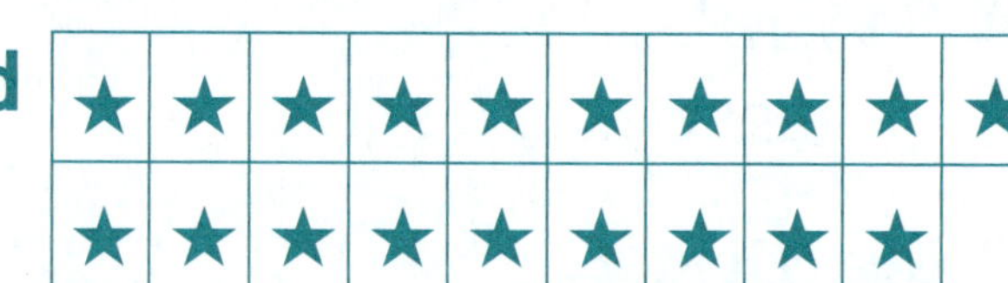

☐

e

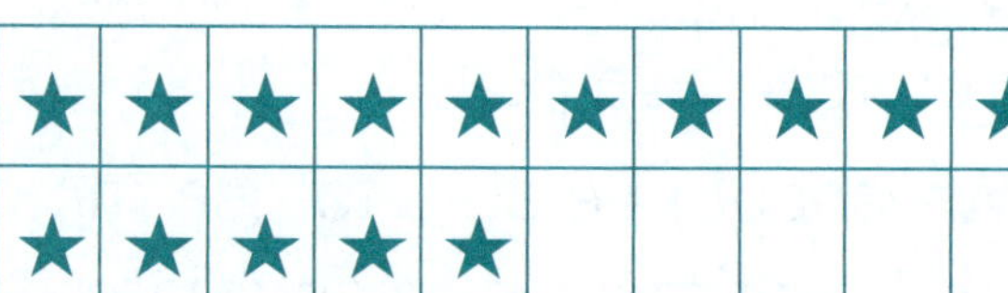

Answers on page A1

C

1 Colour the circles.

a

b

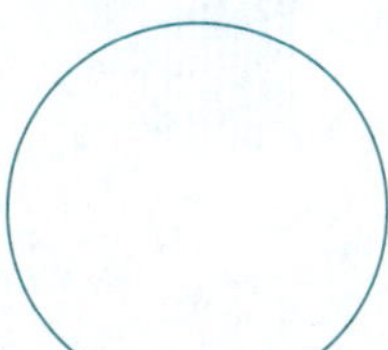

c

d

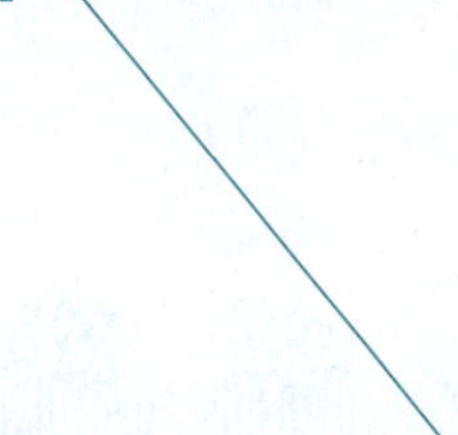

e

f

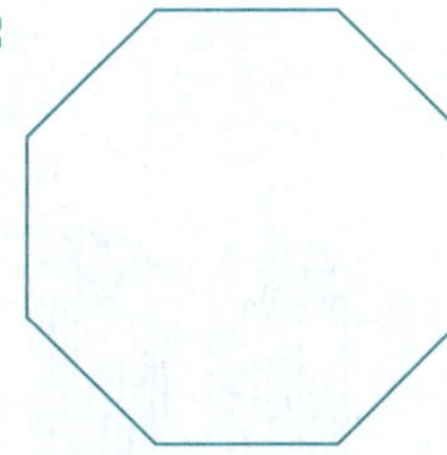

g

h

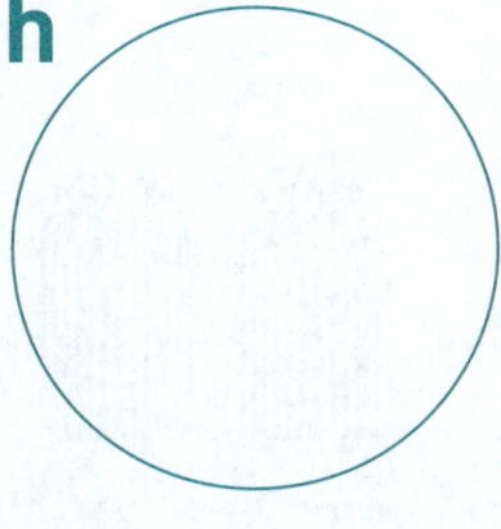

i

D

1 Colour the:

a long snake.

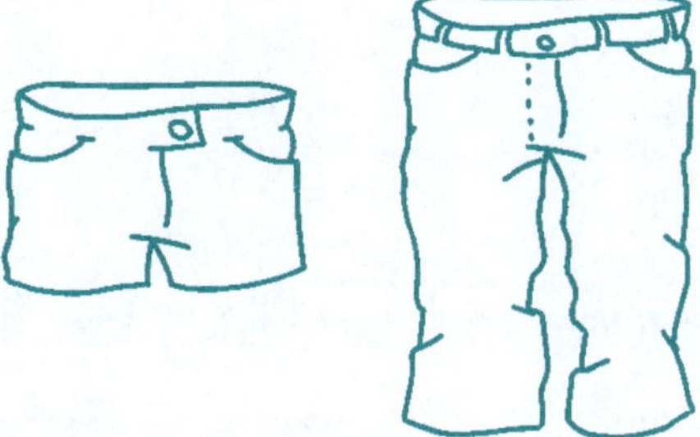

b short trousers.

c tall camel.

d big frog.

Unit 2

A

1 Complete.

a

☐ plus 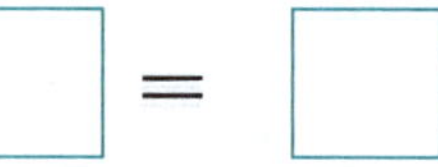

b

☐ plus 

c

☐ plus 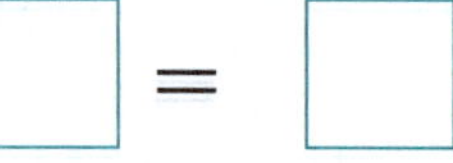

d

☐ plus 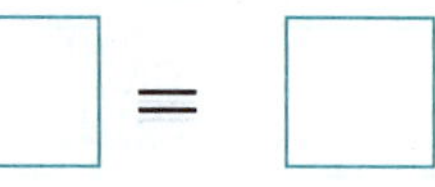

e

☐ plus 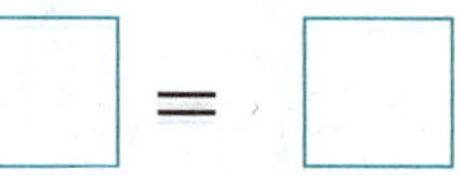

B

1 Write the numeral for:

a  ☐

b  ☐

c  ☐

d  ☐

e   ☐

Answers on page A1

C

1 Draw a curved line.

2 Draw a square.

3 Draw a triangle.

4 Draw a circle.

5 Draw a straight line between these two points.

D

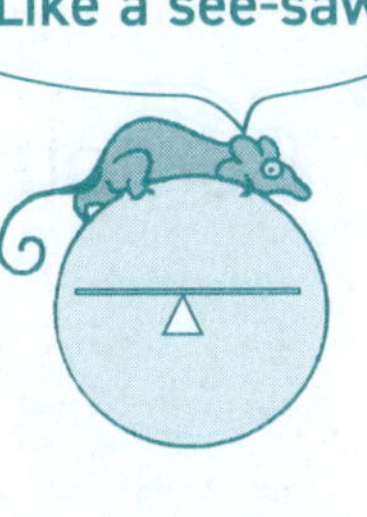

1 Colour the objects that have the same mass.

a

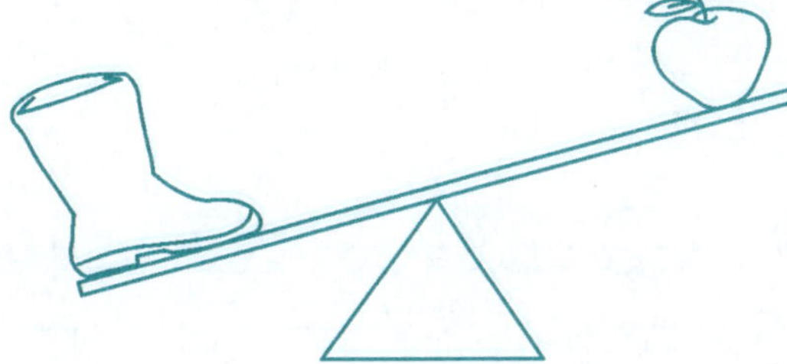

b

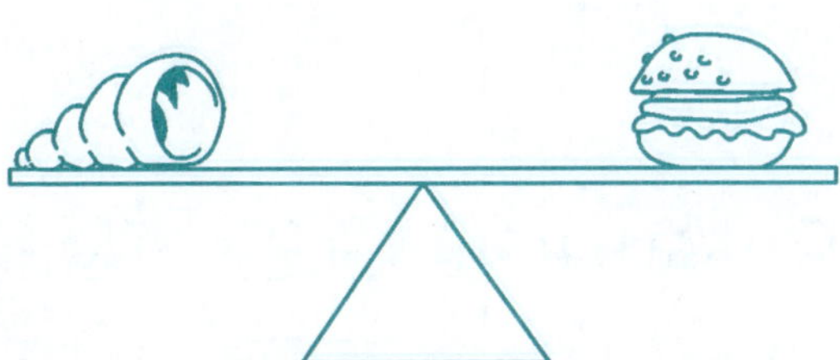

c

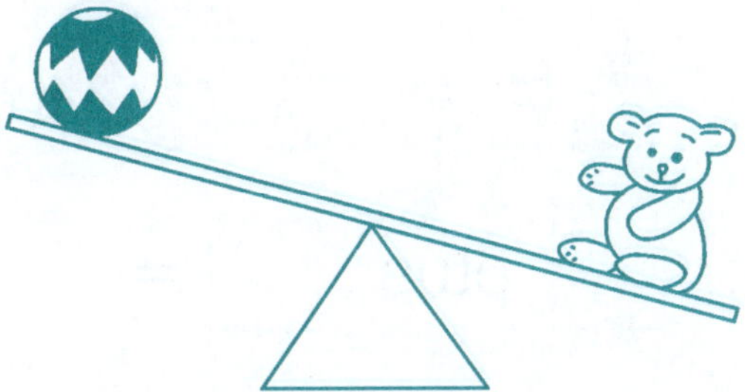

d

Unit 3

A

1 Complete.

a 

☐ plus ☐ = ☐

b 

☐ plus ☐ = ☐

c 

☐ plus ☐ = ☐

d 

☐ plus ☐ = ☐

e 

☐ plus ☐ = ☐

B

2 plus 3 = 5

1 Write the numeral for:

a

☐

b

☐

c

☐

d 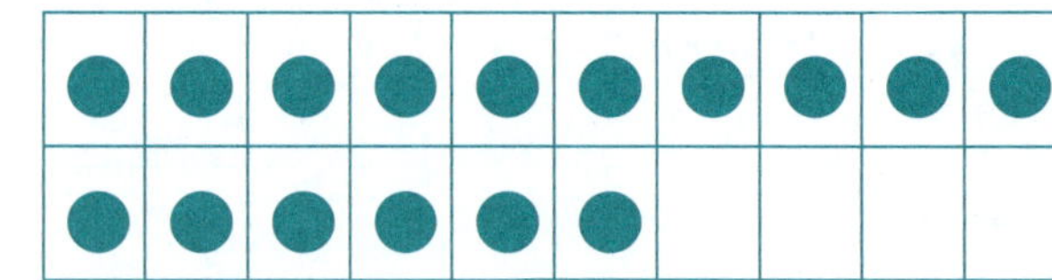

☐

e 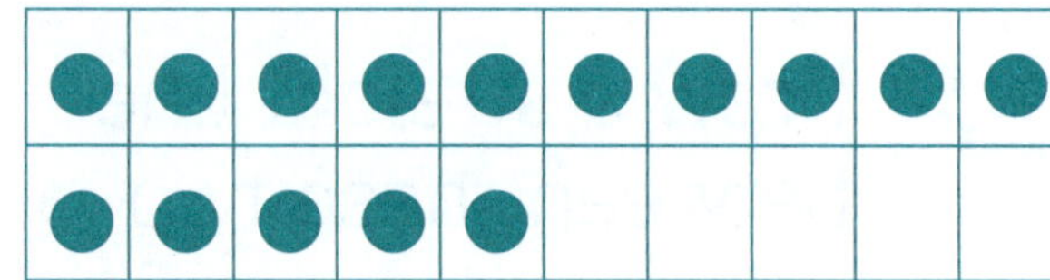

☐

Answers on page A1

C

1 Colour the squares.

a
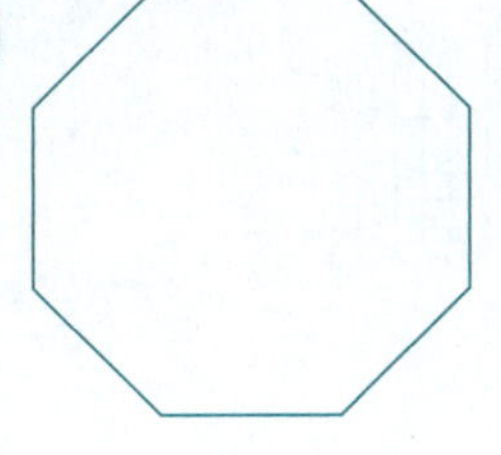

b

c
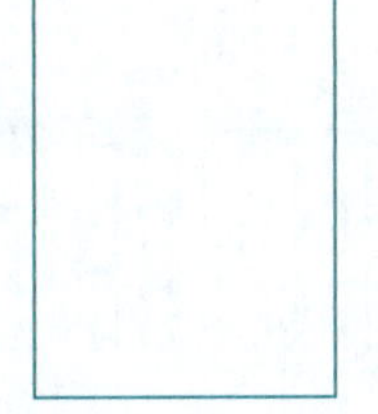

d
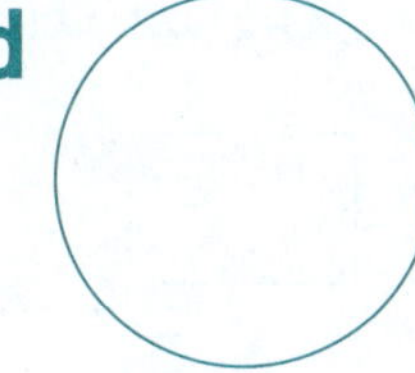

e
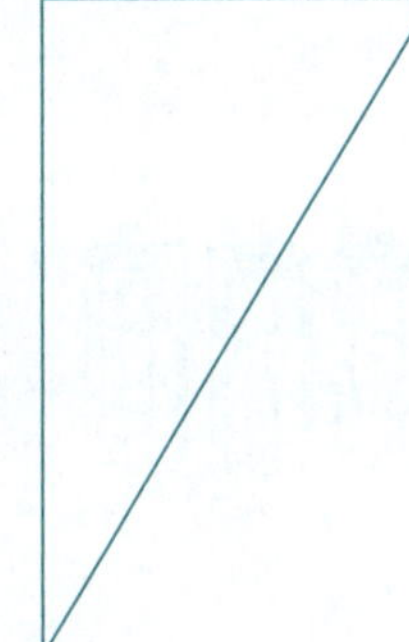

f

g
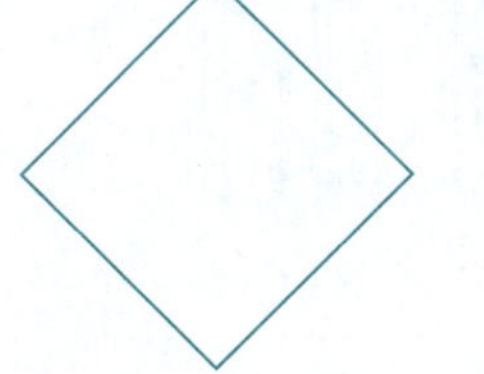

h

i

j
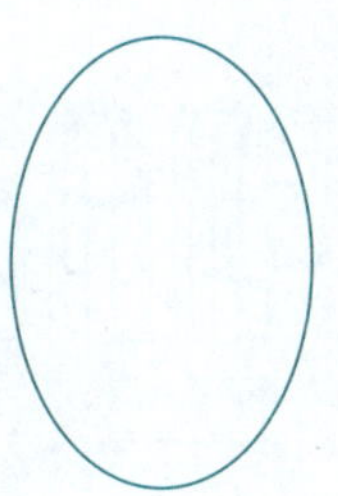

D

1 Colour objects that would be longer than your finger.

a

b

c

d

e

f

g

h
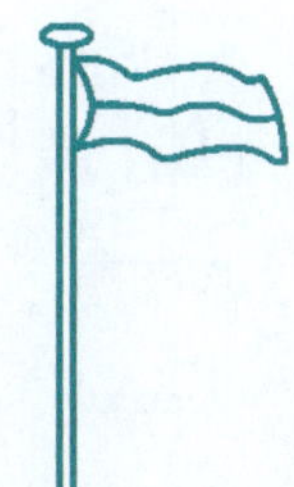

i

Unit 4

A

1 Complete.

a 

☐ + ☐ = ☐

b 

☐ + ☐ = ☐

c

☐ + ☐ = ☐

d

☐ + ☐ = ☐

e

☐ + ☐ = ☐

B

We read this as '2 plus 3'.

2 + 3

1 Write the numeral for:

a 

☐

b 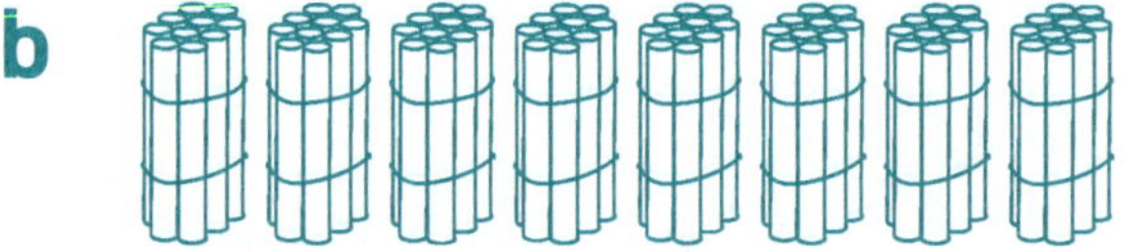

☐

c 

☐

d 

☐

e 

☐

Answers on page A2

C

1 Colour objects that would roll.

a 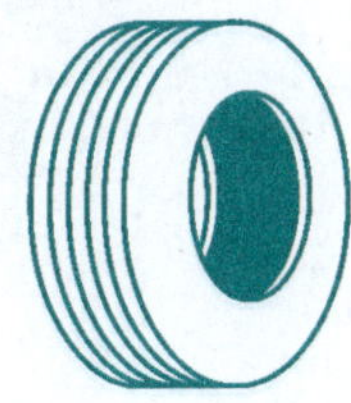

b 

c 

d 

e 

f 

g 

h 

i 

j 

k 

D

This shows 3 o'clock.

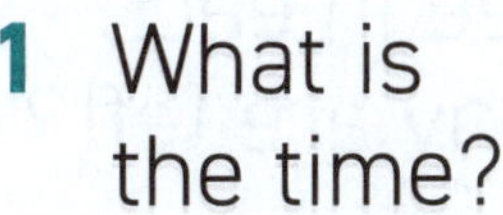

1 What is the time?

a 

______ o'clock

b 

______ o'clock

c 

______ o'clock

d 

______ o'clock

Unit 5

A

1 Cross off three in each row. How many are left?

a 

☐ left

b 

☐ left

c 

☐ left

d 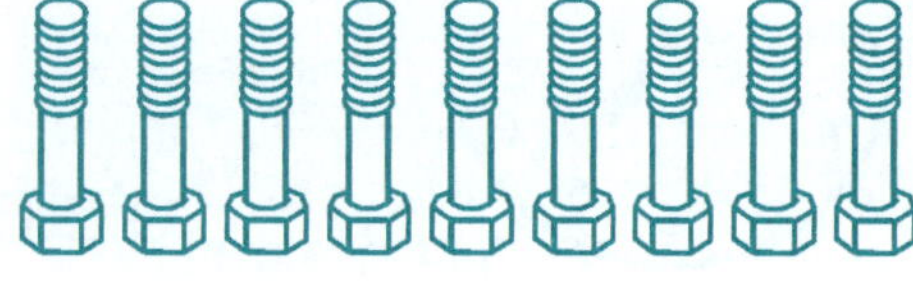

☐ left

e 

☐ left

B

1 out of 2 equal parts

1 Colour the fraction shown.

a  one half

b one quarter

c 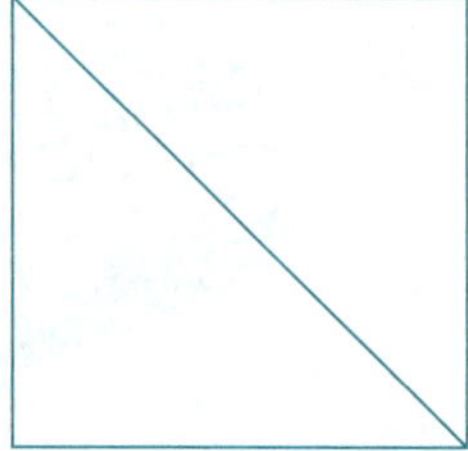 one half

d 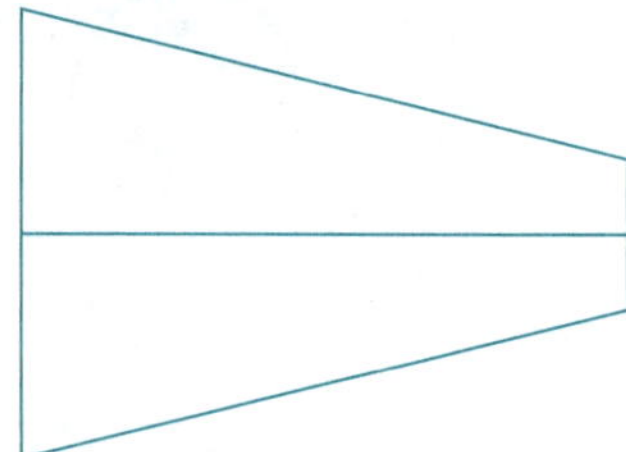 one half

e  one quarter

Answers on page A2

C

1 Colour the rectangles.

a

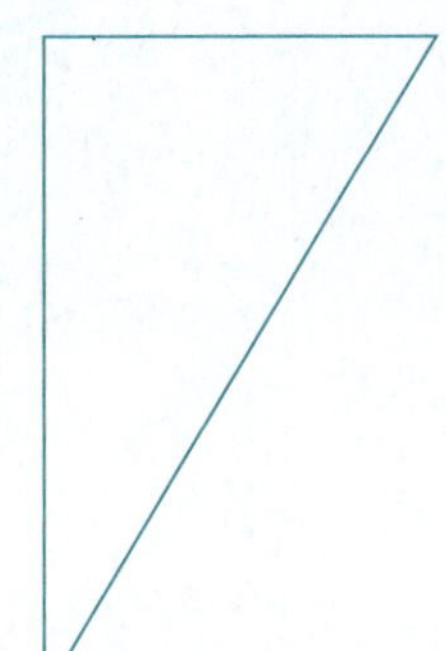

b

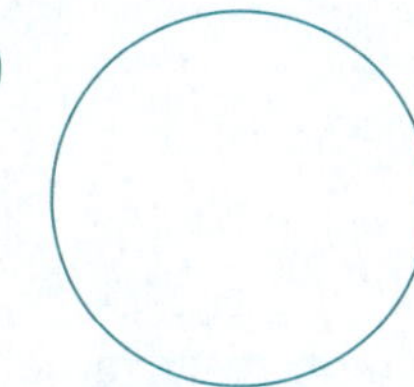

c

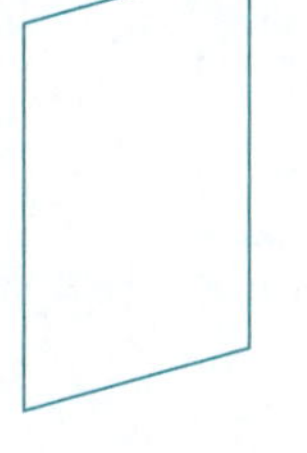

d

e

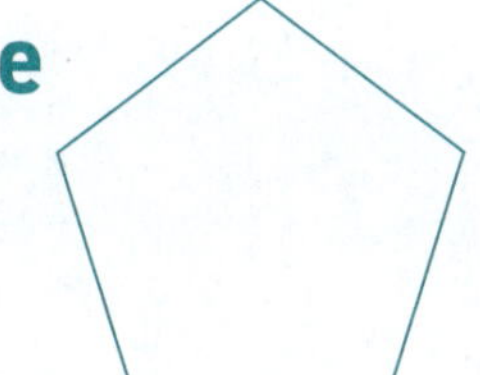

f

g

h

i

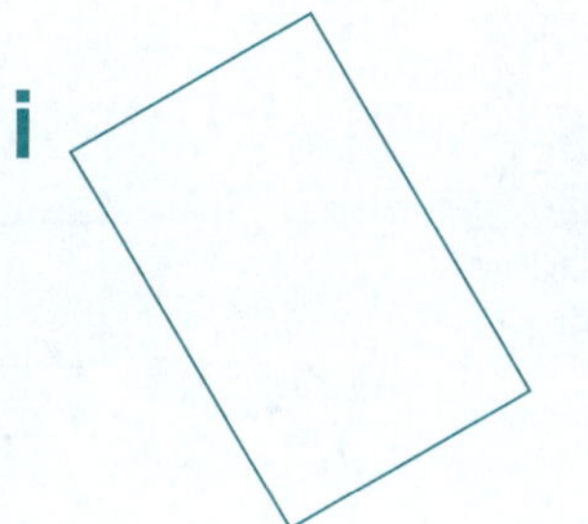

D

1 Colour the objects that have the same mass.

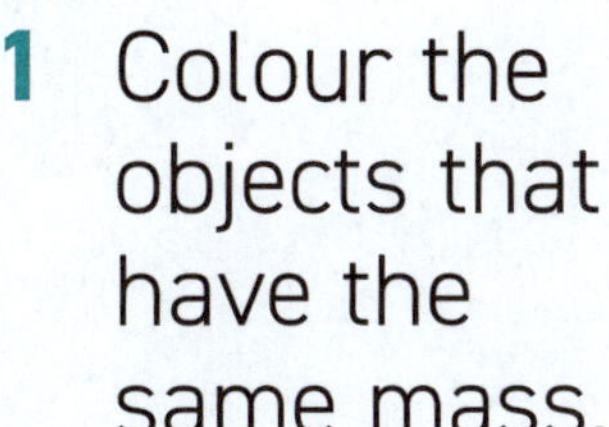

a

b

c

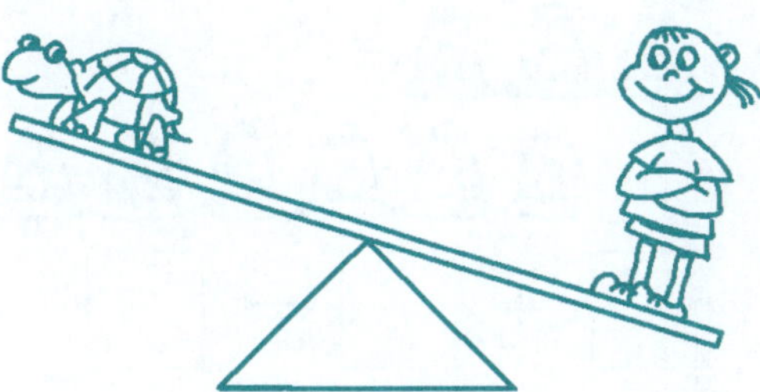

d

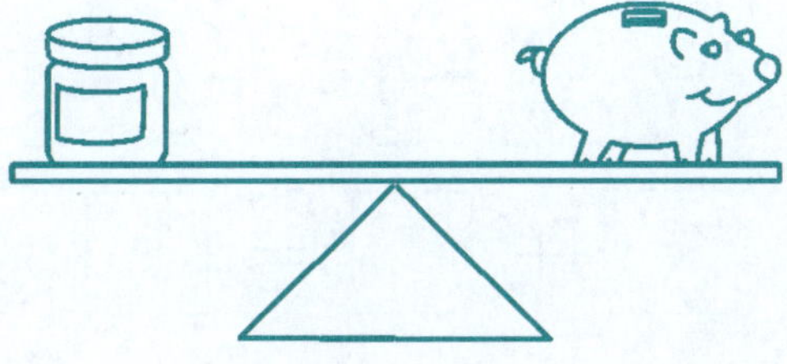

Unit 6

A

1 Complete.

a

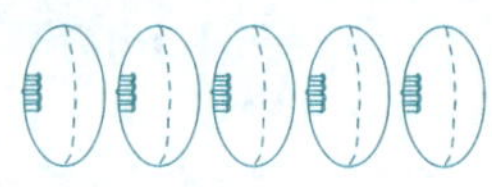

☐ + ☐ = ☐

b

☐ + ☐ = ☐

c

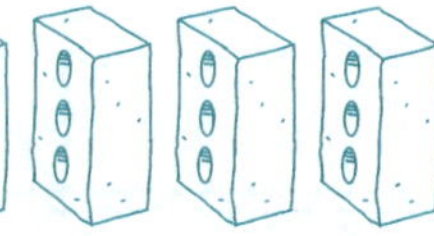

☐ + ☐ = ☐

d

☐ + ☐ = ☐

e

☐ + ☐ = ☐

B

1 Write the numeral for:

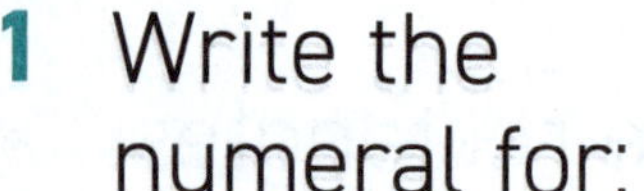

a

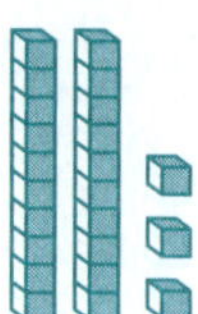

☐

b

☐

c

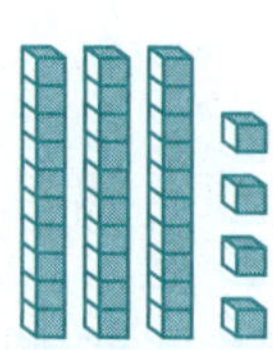

☐

d

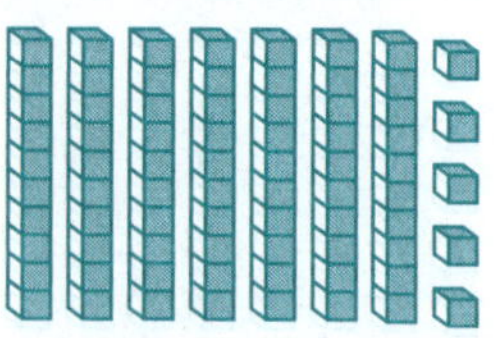

☐

e

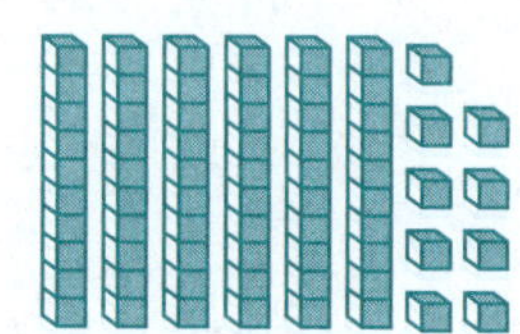

☐

Answers on page A2

C

1 Colour the objects that would stack.

a 

b 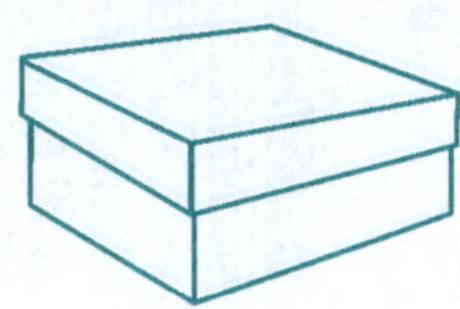

c 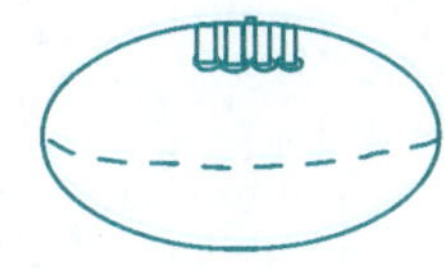

d 

e 

f 

g 

h 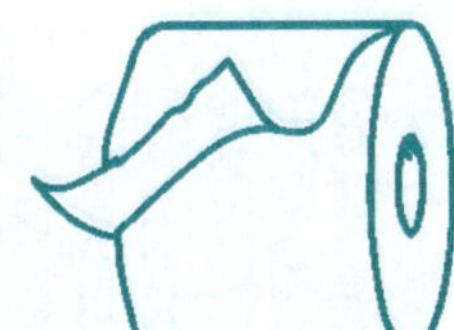

i 

j 

D

1 Colour the larger container.

a  

b 

c  

d  

e  

Unit 7

A

1 Complete.

a   

☐ take away ☐ = ☐

b 

☐ take away ☐ = ☐

c 

☐ take away ☐ = ☐

d 

☐ take away ☐ = ☐

e 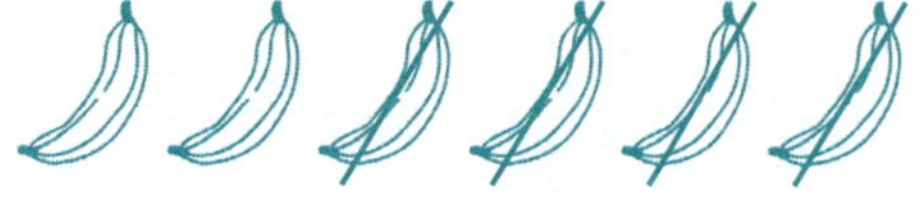

☐ take away ☐ = ☐

B

3 take away 2 = 1

1 Write the numeral for:

a 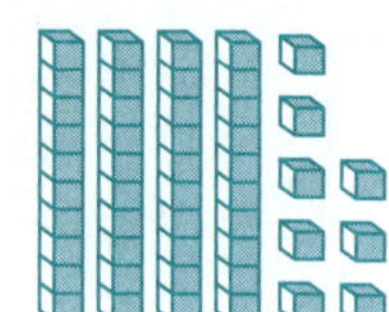 ☐

b 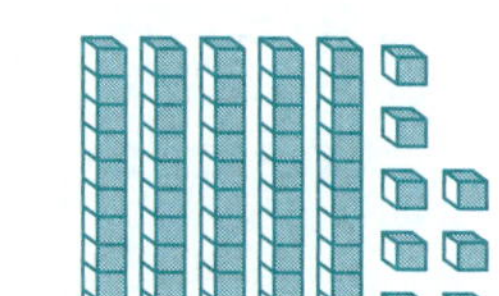 ☐

c 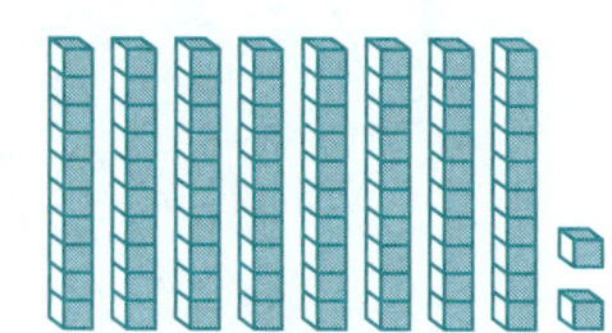 ☐

d 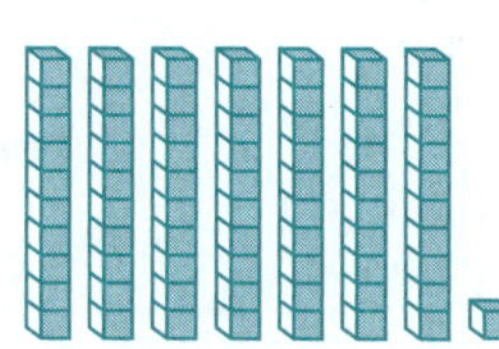 ☐

e  ☐

Answers on page A3

C

1 Colour the triangles.

a
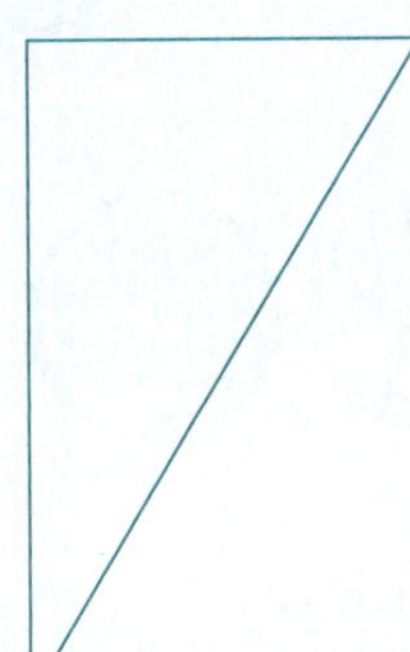

b

c
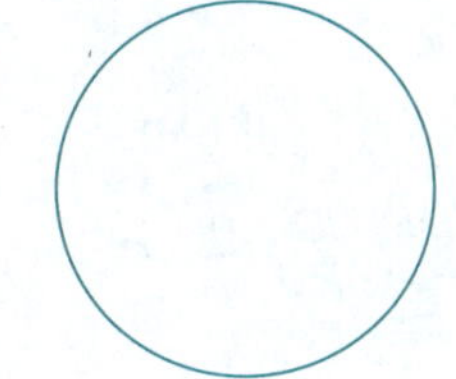

d
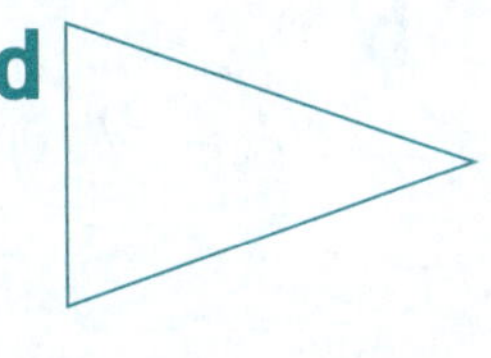

e
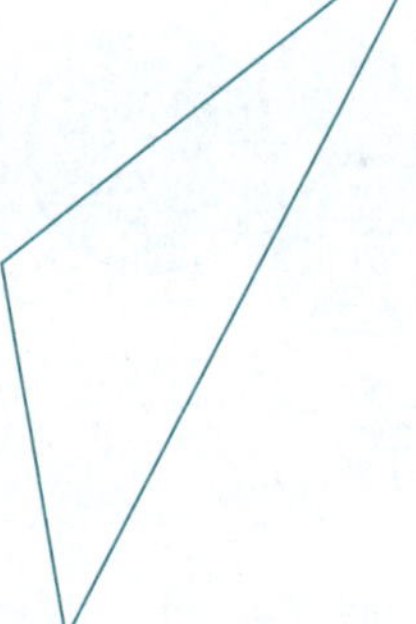

f
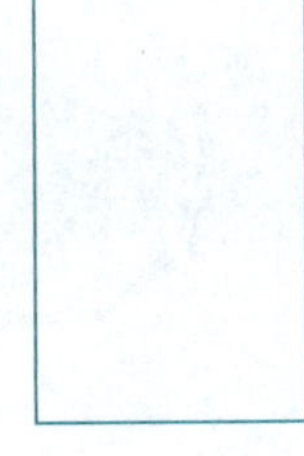

g
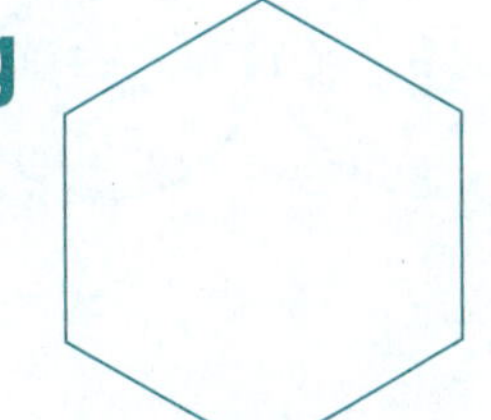

h
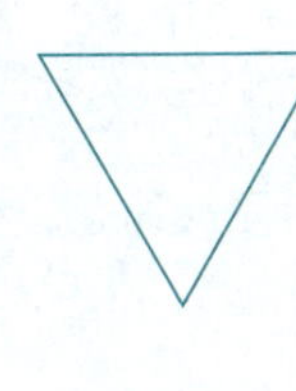

i
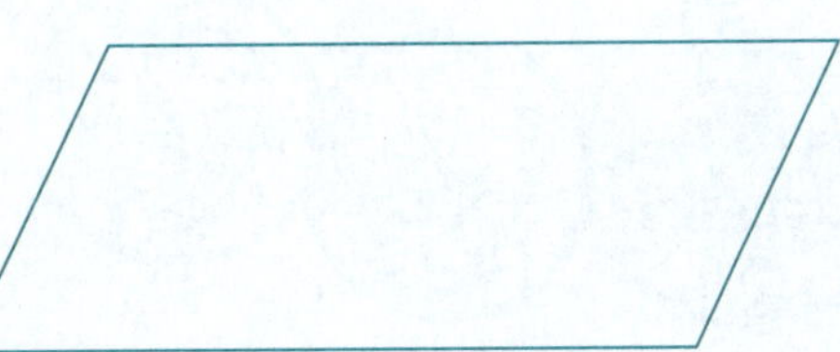

D

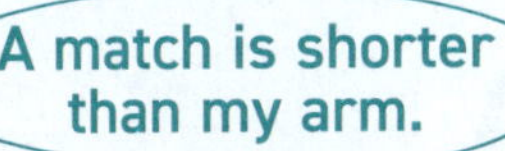

1 Colour objects that would be shorter than your arm.

a

b

c
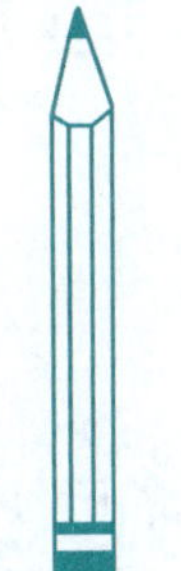

d

e
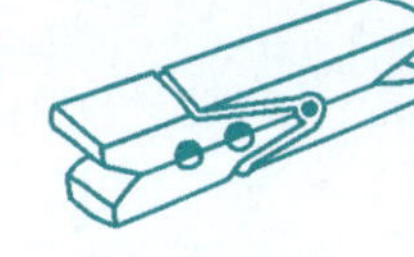

f

g

h

A

1 Complete.

a

☐ take away ☐ = ☐

b

☐ take away ☐ = ☐

c

☐ take away ☐ = ☐

d

☐ take away ☐ = ☐

e

☐ take away ☐ = ☐

B

We write 60c for sixty cents.

1 How much altogether?

a

______ cent

b

______ cent

c

______ cent

d

______ cent

e

______ cent

C

1 Colour the quadrilaterals.

a

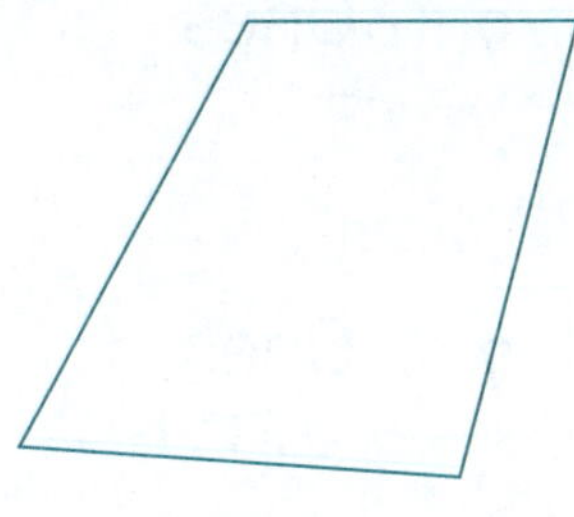

b

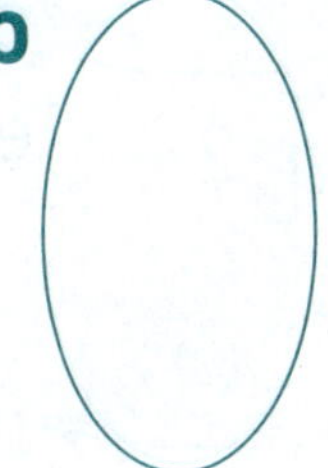

c

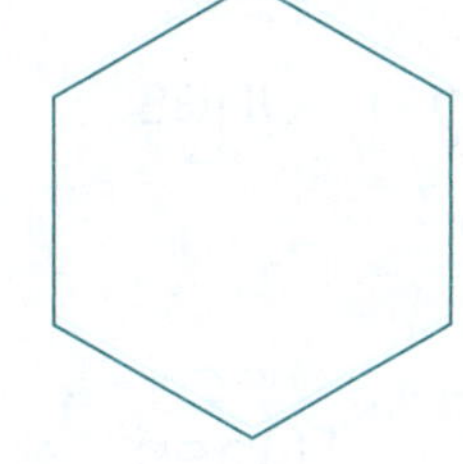

d

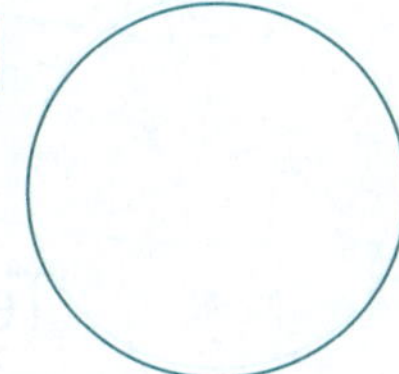

e

f

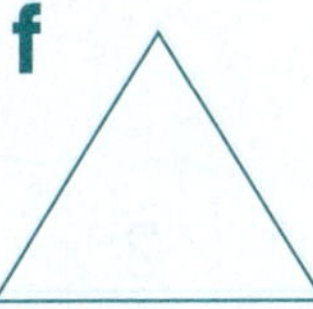

g

h

i

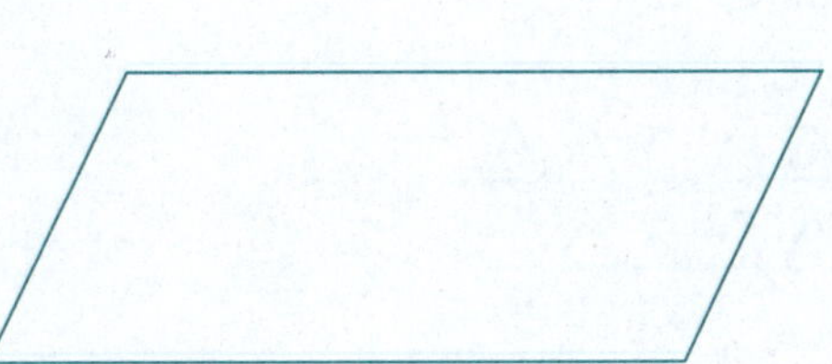

D

1 Colour the heavier object.

a

b

c

d

Unit 9

A

1 Complete.

a $2 + 5 = \square$

b $3 + 4 = \square$

c $4 + 3 = \square$

d $2 + 6 = \square$

e $1 + 5 = \square$

f $5 + 2 = \square$

B

1 Write the numeral.

That's 25.

2 Tens 5 Ones

a 1 Tens 9 Ones ______

b 2 Tens 7 Ones ______

c 4 Tens 2 Ones ______

d 5 Tens 1 Ones ______

e 6 Tens 4 Ones ______

f 3 Tens 6 Ones ______

1 Colour the even groups.

a ○○○○○○○
○○○○○○○

b △△△△△△△△△
△△△△△△△△

c ⬡⬡⬡⬡⬡⬡⬡⬡⬡⬡⬡⬡
⬡⬡⬡⬡⬡⬡⬡⬡⬡⬡⬡⬡

Answers on page A3

C

1 Complete each pattern.

a

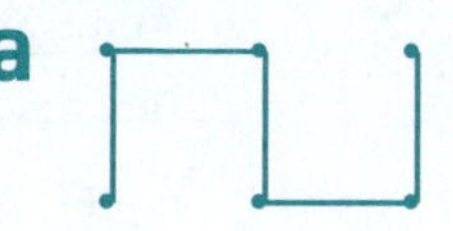

b

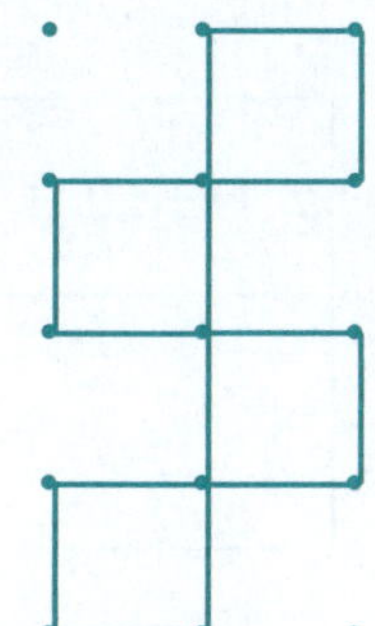

2 How many squares in pattern **b**? ________

D

When the big hand is on 12 it is o'clock.

1 Write the time.

a

________ o'clock

b

half past ________

c

________ o'clock

d

half past ________

Unit 10

A

1 Complete.

a

☐ minus ☐ = ☐

b

☐ minus ☐ = ☐

c

☐ minus ☐ = ☐

d

☐ minus ☐ = ☐

e

☐ minus ☐ = ☐

B

1 Write the numbers.

a

before		after
	26	

b

before		after
	17	

c

before		after
	35	

d

before		after
	29	

e

before		after
	20	

Answers on page A4

C

1 Colour the shapes which have curved sides.

a

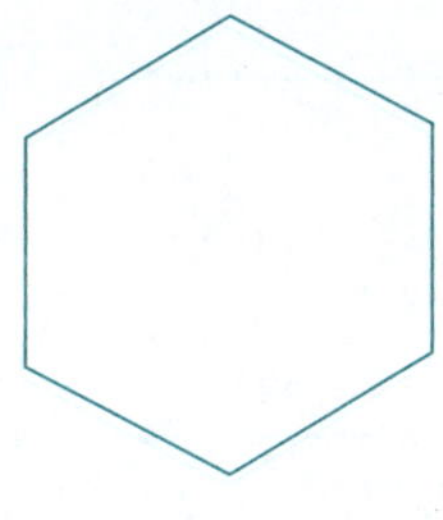

b

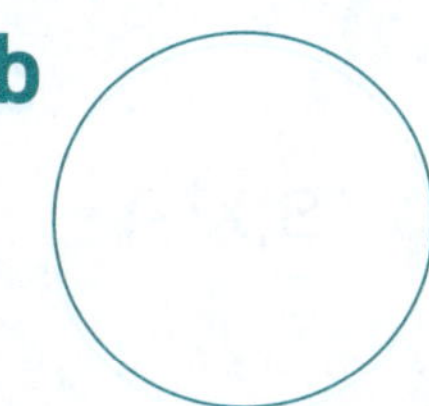

c

d

e

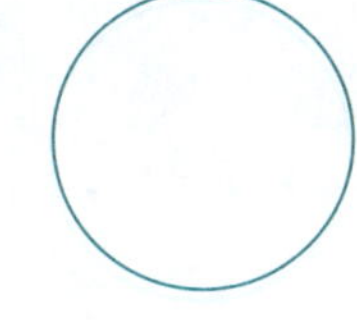

f

g

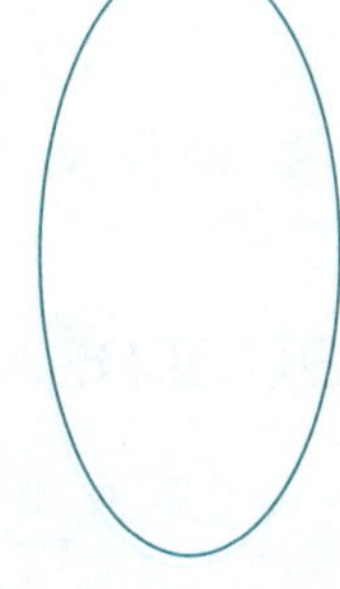

h

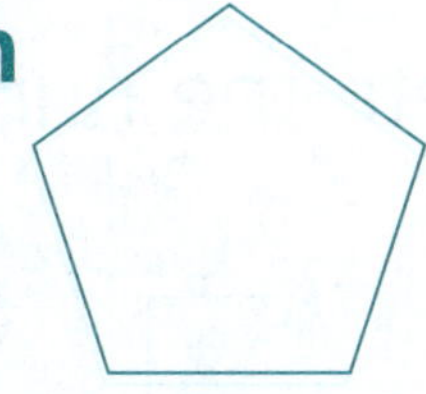

i

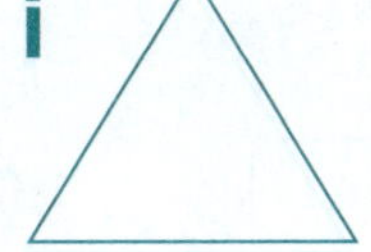

j

D

The oval has a curved side.

1 Colour the larger area.

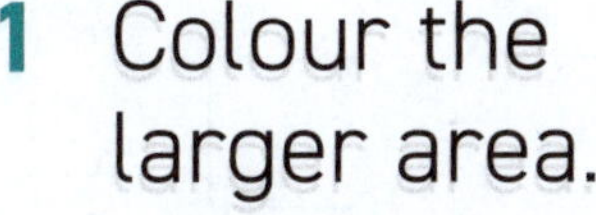

a

b

c

d

Unit 11

A

1 Complete.

a $5 + 1 = \square$

b $6 + 1 = \square$

c $4 + 4 = \square$

d $5 + 3 = \square$

e $2 + 7 = \square$

f $3 + 5 = \square$

B

This says fift

5th

1 Write the ordinal number for:

a third ____________

b sixth ____________

c second ____________

d tenth ____________

e fourth ____________

f first ____________

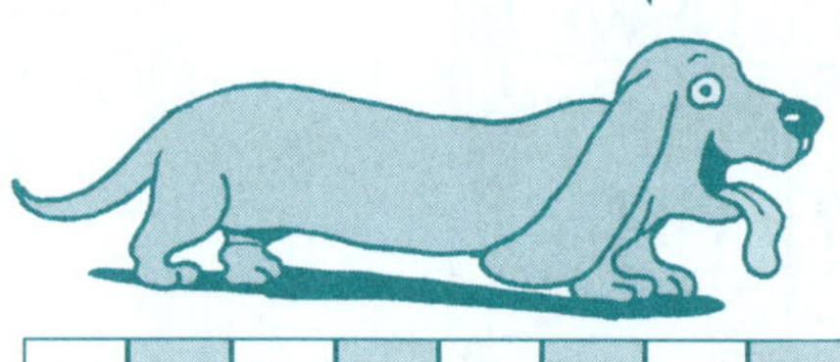

1 Complete the number patterns.

a

2 , 4 , 6 , □

b

10 , 12 , 14 , □

C

1 Colour the hexagons.

a

b

c

d

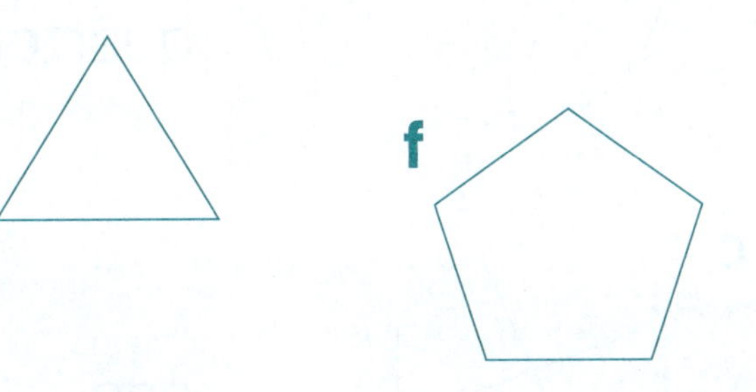

e

f

g

h

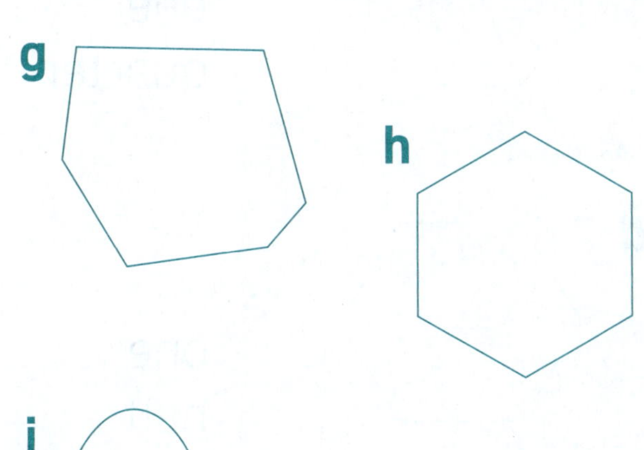

i

j

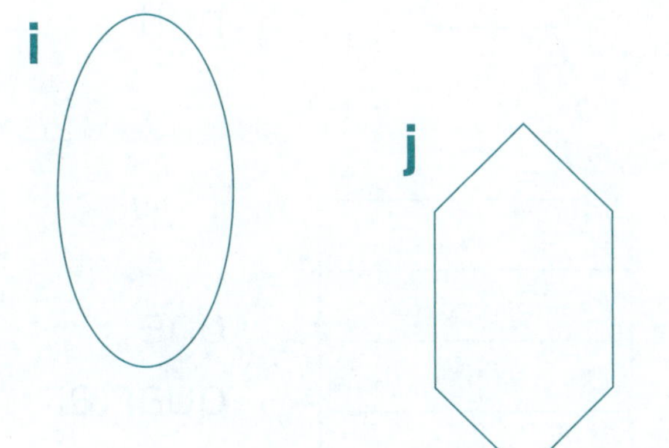

D

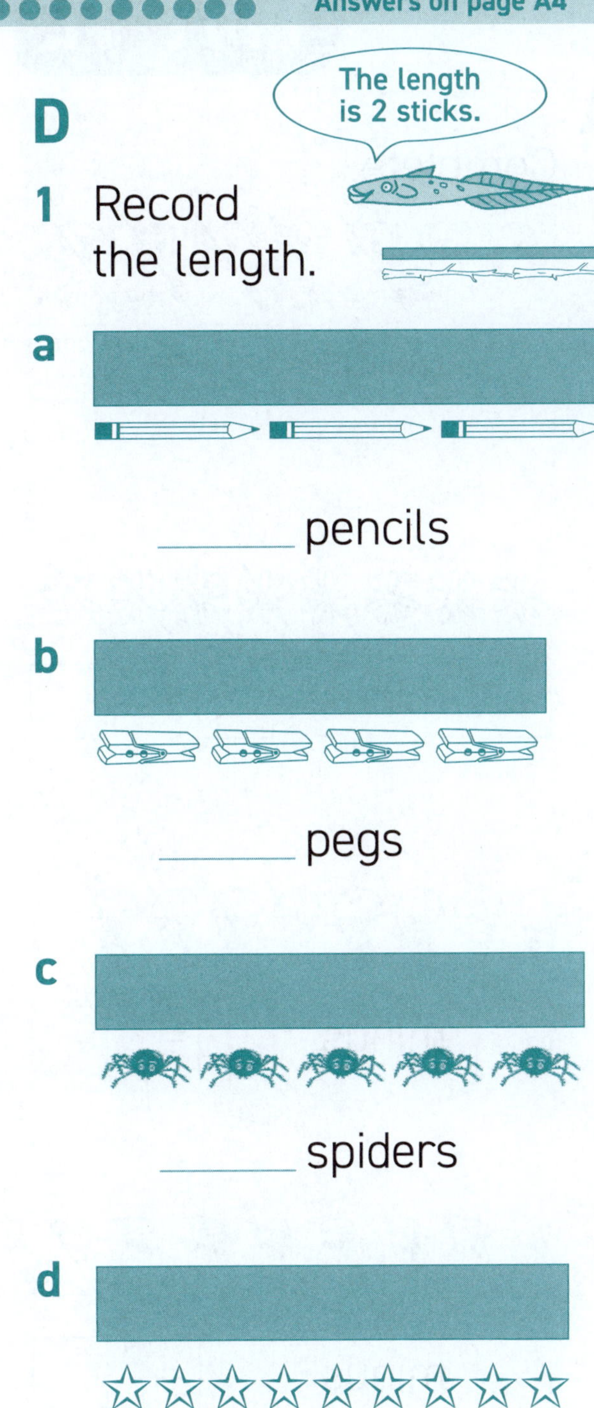

1 Record the length.

a

_____ pencils

b

_____ pegs

c

_____ spiders

d

_____ stars

e

_____ dinosaurs

Unit 12

A

1 Complete.

a

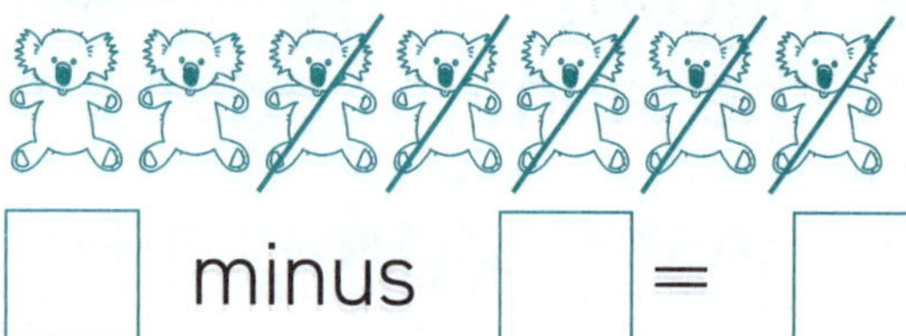

☐ minus ☐ = ☐

b

☐ minus ☐ = ☐

c

☐ minus ☐ = ☐

d

☐ minus ☐ = ☐

e

☐ minus ☐ = ☐

B

1 Colour the fraction shown.

a

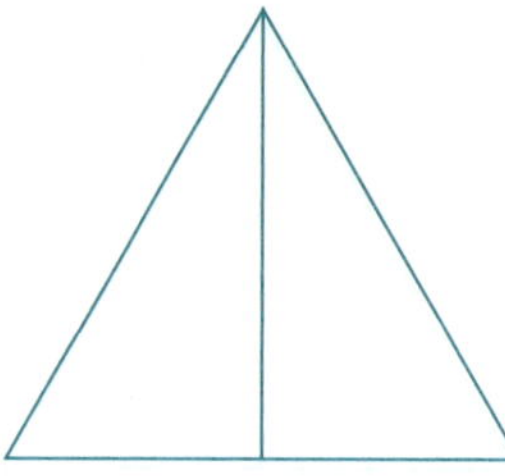

one half

b

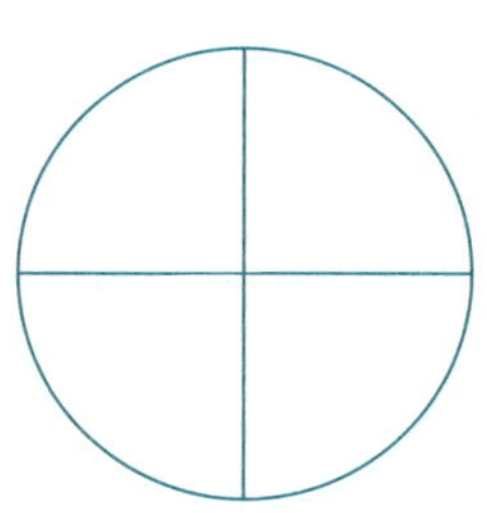

one quarter

c

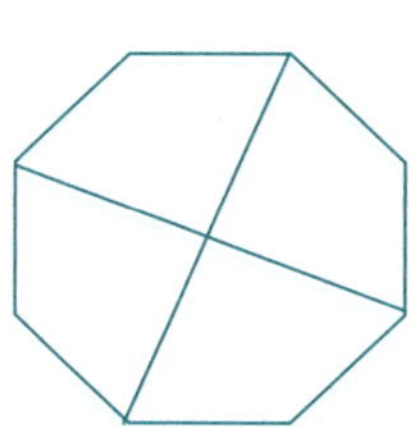

one quarter

d

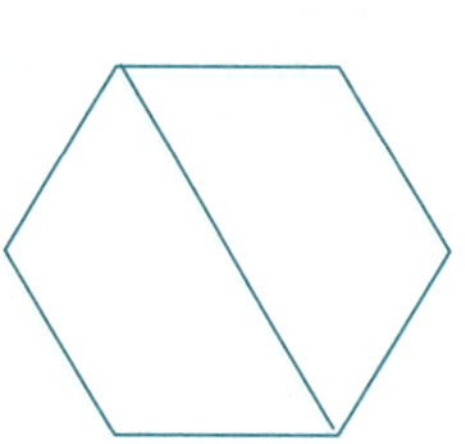

one half

e

one quarter

Answers on page A4

C

1 Complete.

	Shape	Number of sides
a	▲	
b	■	
c	▮	
d	⬢	
e	⯃	

D

1 Colour the lighter object.

a

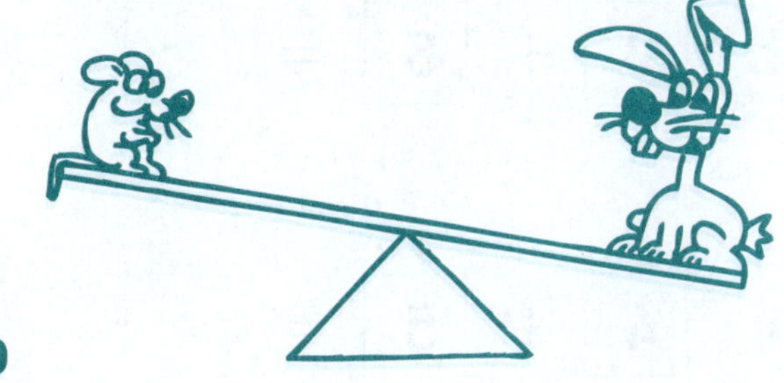

b

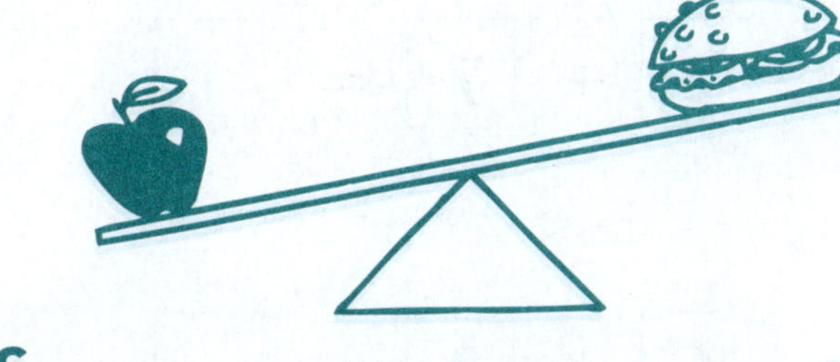

c

d

Unit 13

A

1 Complete.

a 3 + 6 = ☐

b 2 + 8 = ☐

c 4 + 5 = ☐

d 2 + 7 = ☐

e 6 + 2 = ☐

f 7 + 1 = ☐

B

1 Write the numeral.

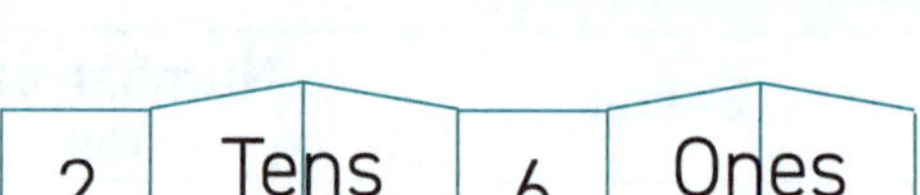

a 2	Tens	6	Ones	______
b 6	Tens	3	Ones	______
c 8	Tens	5	Ones	______
d 4	Tens	7	Ones	______
e 6	Tens	2	Ones	______
f 3	Tens	8	Ones	______

C

1 Complete the number patterns.

a

b 12 , 11 , 10 , ☐

Unit 1 .... Page 18

A **1 a** 6 **b** 7 **c** 8 **d** 8 **e** 9

B **1 a** 16 **b** 17 **c** 18 **d** 19 **e** 15

C **1** Parts b, e, h and i are coloured.

D **1 a** Top long snake is coloured.
b Left short trousers are coloured.
c Right tall camel is coloured.
d Left big frog is coloured.

Unit 2 .... Page 20

A **1 a** 4 plus 3 = 7 **b** 3 plus 5 = 8 **c** 5 plus 3 = 8 **d** 3 plus 3 = 6 **e** 2 plus 7 = 9

B **1 a** 10 **b** 30 **c** 20 **d** 50 **e** 40

C **1** Check with teacher or parent
2 Check with teacher or parent
3 Check with teacher or parent
4 yes
5 Check with teacher or parent

D **1 a** are not coloured **b** are coloured **c** are not coloured **d** are coloured.

Unit 3 .... Page 22

A **1 a** 2 plus 4 = 6 **b** 3 plus 3 = 6 **c** 5 plus 2 = 7 **d** 4 plus 3 = 7 **e** 3 plus 5 = 8

B **1 a** 12 **b** 14 **c** 13 **d** 16 **e** 15

C **1** Parts b, f, g and i are coloured.

D **1** Parts a, c, d, f, g and h are coloured.

Unit 4. . . . . Page 24

A **1 a** 4 + 3 = 7 **b** 6 + 1 = 7 **c** 5 + 2 = 7 **d** 1 + 6 = 7 **e** 5 + 4 = 9

B **1 a** 50 **b** 80 **c** 70 **d** 60 **e** 90

C **1** Parts a, d, e, h, i and j are coloured.

D **1 a** 2 o'clock **b** 8 o'clock **c** 5 o'clock **d** 11 o'clock

Unit 5. . . . . Page 26

A **1 a** 2 **b** 3 **c** 5 **d** 6 **e** 4

B **1 a** **b** **c** **d** **e**

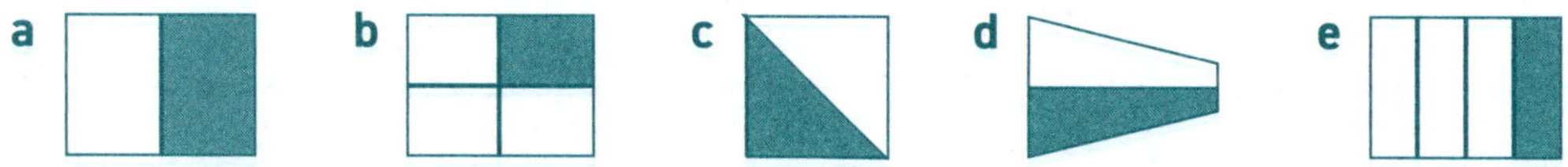

C **1** Parts d, f, g and i are coloured.

D **1 a** are coloured **b** are coloured **c** are not coloured **d** are coloured

Unit 6. . . . . Page 28

A **1 a** 5 + 5 = 10 **b** 6 + 3 = 9 **c** 4 + 3 = 7 **d** 3 + 7 = 10 **e** 6 + 4 = 10

B **1 a** 23 **b** 43 **c** 34 **d** 75 **e** 69

C **1** Parts a, b, d, e, g, h and i are coloured.

D **1 a** kettle is coloured **b** bucket is coloured **c** bin is coloured **d** jug is coloured
e jug is coloured

Unit 7 . . . . Page 30

A **1 a** 5 take away 2 = 3 **b** 6 take away 3 = 3 **c** 4 take away 2 = 2 **d** 7 take away 3 = 4
e 6 take away 4 = 2

B **1 a** 48 **b** 58 **c** 82 **d** 71 **e** 96

C **1** Parts a, d, e and h are coloured.

D **1** Parts a, c, d, e, f and g are coloured.

Unit 8 . . . . Page 32

A **1 a** 6 take away 5 = 1 **b** 5 take away 3 = 2 **c** 5 take away 4 = 1 **d** 6 take away 4 = 2
e 8 take away 3 = 5

B **1 a** 50c **b** 50c **c** 80c **d** 80c **e** 90c

C **1** Parts a, e, g and i are coloured.

D **1 a** hen is coloured **b** dog is coloured **c** rooster is coloured **d** boy is coloured

Unit 9. . . . . Page 34

A **1 a** 7 **b** 7 **c** 7 **d** 8 **e** 6 **f** 7

B **1 a** 19 **b** 27 **c** 42 **d** 51 **e** 64 **f** 36

Extra practice section: 1 a is coloured **b** is not coloured **c** is coloured

C **1 a** **b** **2** 15

D 1 **a** 1 o'clock **b** half past 2 **c** 3 o'clock **d** half past 10

Unit 10 .... Page 36

A 1 **a** 6 minus 2 = 4 **b** 5 minus 1 = 4 **c** 8 minus 2 = 6 **d** 4 minus 3 = 1
e 7 minus 4 = 3

B 1 **a** 25, 26, 27 **b** 16, 17, 18 **c** 34, 35, 36 **d** 28, 29, 30 **e** 19, 20, 21

C 1 Parts b, d, e, g and j are coloured.

D 1 **a** 75c stamp is coloured **b** ice cream lid is coloured **c** 50c coin is coloured
d newspaper is coloured

Unit 11 .... Page 38

A 1 **a** 6 **b** 7 **c** 8 **d** 8 **e** 9 **f** 8

B 1 **a** 3rd **b** 6th **c** 2nd **d** 10th **e** 4th **f** 1st

Extra practice section: 1 a 2, 4, 6, 8 **b** 10, 12, 14, 16

C 1 Parts b, d, g, h and j are coloured.

D 1 **a** 3 pencils **b** 4 pegs **c** 5 spiders **d** 9 stars **e** 5 dinosaurs

Unit 12 .... Page 40

A 1 **a** 7 minus 5 = 2 **b** 8 minus 5 = 3 **c** 8 minus 4 = 4 **d** 9 minus 3 = 6
e 9 minus 5 = 4

B 1 **a** **b** 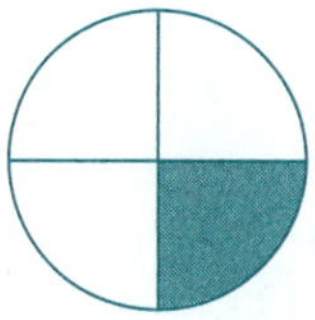 **c**

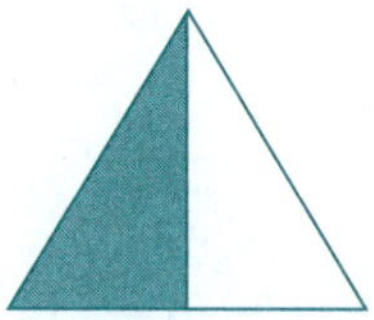

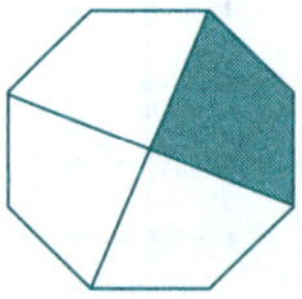

d 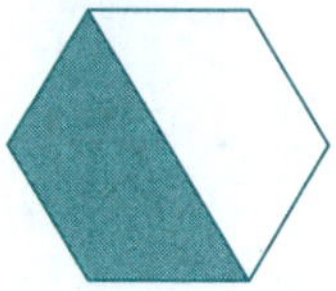 **e**

C 1 **a** 3 sides **b** 4 sides **c** 4 sides **d** 6 sides **e** 8 sides

D 1 **a** mouse is coloured **b** hamburger is coloured **c** snail is coloured
d tortoise is coloured

Unit 13 .... Page 42

A **1 a** 9 **b** 10 **c** 9 **d** 9 **e** 8 **f** 8

B **1 a** 26 **b** 63 **c** 85 **d** 47 **e** 62 **f** 38

Extra practice section: 1 a 6, 5, 4, 3 **b** 12, 11, 10, 9

C **1** 1 circle, 3 triangles, 3 hexagons, 2 rectangles

D **1 a** orange is coloured **b** car is coloured **c** die is coloured **d** cap is coloured
e small dinosaur on the left is coloured

Unit 14 .... Page 44

A **1 a** 4 − 3 = 1 **b** 7 − 5 = 2 **c** 5 − 4 = 1 **d** 9 − 4 = 5 **e** 6 − 4 = 2

B **1 a** 29, 30, 31 **b** 18, 19, 20 **c** 23, 24, 25 **d** 30, 31, 32 **e** 47, 48, 49

C **1 a** yes **b** no **c** yes **d** no **e** no **f** yes

D **1 a** **b**  **c**  **d**  **e**

Unit 15 .... Page 46

A **1 a** 9 **b** 9 **c** 10 **d** 9 **e** 9 **f** 10

B **1 a** 25 **b** 16 **c** 31 **d** 19 **e** 30 **f** 43

Extra practice section: 1 a is coloured **b** is not coloured **c** is coloured

C **1 a** 4 **b** 3 **c** 2
2 a yes **b** no

D **1 a** 3 blocks **b** 5 blocks **c** 4 blocks **d** 5 blocks **e** 4 blocks

Unit 16 .... Page 48

A 1 **a** 2 rows of 4 = 8 **b** 2 rows of 6 = 12 **c** 2 rows of 3 = 6 **d** 2 rows of 7 = 14
e 2 rows of 5 = 10

B 1 **a** a 20 and 10 cent coin coloured **b** a 20, a 10 and a 5 cent coin coloured
c a 50, a 10 and a 5 cent coin coloured **d** two 20 and a 10 cent coin coloured
e a 50 and a 20 cent coin coloured

C 1 Parts b, e and h are coloured.

D 1 **a** 4 forks **b** 4 pencils **c** 5 sticks **d** 7 stars **e** 7 blocks

Unit 17 .... Page 50

A 1 **a** 6 − 2 = 4 **b** 5 − 2 = 3 **c** 7 − 6 = 1 **d** 5 − 3 = 2 **e** 9 − 6 = 3

B 1 **a** 7 **b** 7 **c** 20 **d** 20 **e** 10

C 1 Parts c, f and i are coloured.

D 1 Sunday, Monday, Tuesday, Wednesday, Thursday, Friday, Saturday

Unit 18 .... Page 52

A 1 **a** 2 **b** 2 **c** 5 **d** 4 **e** 7 **f** 1

B 1 **a** 32 **b** 15 **c** 28 **d** 53 **e** 37 **f** 61

Extra practice section: 1 a 5th **b** 9th **c** 7th

C 1 **a** peaches are coloured **b** ice cream is coloured **c** soccer ball is coloured
d block is coloured **e** box is coloured

D 1 **a** October **b** June **c** April **d** December **e** February **f** August

Unit 19 .... Page 54

A **1 a** 2 rows of 9 = 18 **b** 3 rows of 4 = 12 **c** 3 rows of 3 = 9 **d** 3 rows of 6 = 18
e 2 rows of 10 = 20

B **1 a**  **b**  **c** 

d  **e** 

C **1** Parts c, f, g, i and j are coloured.

D **1 a**  **b**  **c**  **d**  **e** 

Unit 20 .... Page 56

A **1 a** 5 **b** 3 **c** 7 **d** 3 **e** 6 **f** 4

B **1 a**

1	Tens	4	Ones

b

3	Tens	9	Ones

c

2	Tens	1	Ones

d

5	Tens	6	Ones

e

4	Tens	5	Ones

f

7	Tens	3	Ones

Extra practice section: 1 a is not coloured **b** is coloured **c** is coloured

C **1** 1 square, 2 ovals, 2 octagons, 4 hexagons

D **1 a** 7 rectangles **b** 4 triangles **c** 8 squares **d** 6 rectangles

Unit 21 .... Page 58

A **1 a** 8 **b** 10 **c** 10 **d** 10 **e** 10 **f** 10

B **1 a** 16 **b** 17 **c** 17 **d** 25 **e** 4

Extra practice section: 1 100c (or $1)

C **1 a** 6 **b** 3 **c** 4 **d** 4 **e** 4 **f** 5 **g** 4 **h** 5
2 c, d, e and g are coloured.

D **1 a** March **b** January **c** May **d** September **e** July **f** November

Unit 22 .... Page 60

A **1 a** 2 groups of 3 = 6 **b** 3 groups of 2 = 6 **c** 2 groups of 5 = 10 **d** 2 groups of 6 = 12
e 3 groups of 4 = 12

B **1** **a** **b** 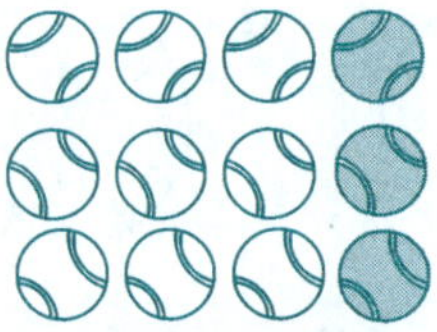 **c**

d  **e**

apple juice apple juice apple juice apple juice
apple juice apple juice apple juice apple juice

C **1 a** 5 **b** 2 **c** 4 **d** 6
2 a yes **b** no

D **1 a** 4 blocks **b** 4 blocks **c** 5 blocks **d** 6 blocks

Unit 23 .... Page 62

A **1 a** 2 **b** 4 **c** 6 **d** 1 **e** 0 **f** 2

B **1 a** 24 **b** 59 **c** 32 **d** 87 **e** 68 **f** 46

Extra practice section: 1 a 7 **b** 3 **c** 4

C 1 a 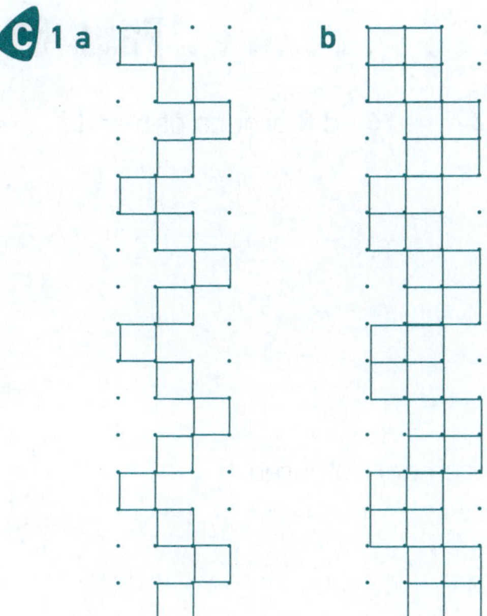

2 a 16 b 32

D 1 a 1 o'clock b 4 o'clock c 10 o'clock d 12 o'clock e 8 o'clock

Unit 24 .... Page 64

A 1 a yes b no c yes d yes e yes

B 1 a 35, 36, 37 b 27, 28, 29 c 52, 53, 54 d 66, 67, 68 e 38, 39, 40

C 1 Refer to the teacher 2 Refer to the teacher 3 Refer to the teacher
4 Refer to the teacher 5 Refer to the teacher

D 1 a 18 squares b 3 rectangles c 6 triangles d 6 rectangles

Unit 25.................................... Page 66

A 1 a 11 b 12 c 11 d 11 e 12 f 11

B 1 a 11th b 15th c 12th d 17th e 16th f 20th

Extra practice section: 1 a 8 **b** 6 **c** 2

C 1 3 sides (4 triangles), 4 sides (square, rectangle and 2 diamonds), 6 sides (2 hexagons)

D 1 a 3 o'clock b 9 o'clock c 11 o'clock d 6 thirty e 5 thirty

Unit 26 . . . . Page 68

A **1** **a** 3 groups of 3 = 9 **b** 2 groups of 3 = 6 **c** 2 groups of 7 = 14 **d** 3 groups of 5 = 15
e 2 groups of 8 = 16

B **1** **a** 45 **b** 74 **c** 34 **d** 67 **e** 88

C **1** **a** 4 **b** 1 **c** 3 **d** 4 **e** 6 **f** 5 **g** 4 **h** 4
2 a, d, g and h are coloured.

D **1** **a** January not coloured **b** February not coloured **c** March not coloured
d April coloured **e** May not coloured **f** June coloured

Unit 27 . . . . Page 70

A **1** **a** 5 **b** 0 **c** 0 **d** 1 **e** 3 **f** 0

B **1** **a** 45 **b** 63 **c** 58 **d** 27 **e** 73 **f** 81

Extra practice section: 1 **a** 3 **b** 3

C **1** **a** 3 **b** 4 **c** 4 **d** 4 **e** 6

D **1** **a** July coloured **b** August coloured **c** September not coloured **d** October coloured
e November not coloured **f** December coloured

Unit 28 . . . . Page 72

A **1** **a** yes **b** yes **c** yes **d** no **e** no

B **1** **a** 42, 43, 44 **b** 68, 69, 70 **c** 46, 47, 48 **d** 75, 76, 77 **e** 54, 55, 56

C **1** Teacher or parent to check.
2 Teacher or parent to check.

D **1** **a** not coloured **b** coloured **c** coloured **d** coloured **e** not coloured **f** not coloured
g not coloured

Unit 29 .... Page 74

A **1 a** 6 **b** 12 **c** 8 **d** 4 **e** 10 **f** 14

B **1 a**  **b** 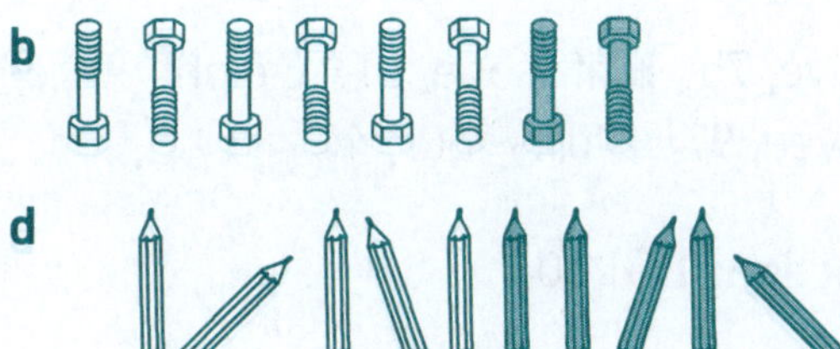

c  **d**

e

Extra practice section: 1 a 4 **b** 4

C **1 a** cylinder **b** cube **c** cone **d** prism **e** sphere

D **1 a** half past 1 **b** half past 8 **c** quarter past 1 **d** half past 7 **e** quarter past 7

Unit 30 .... Page 76

A **1 a** 4 **b** 0 **c** 3 **d** 2 **e** 0 **f** 1

B **1 a** forty-eight, 48 **b** sixty-five, 65 **c** ninety-three, 93 **d** thirty-six, 36
e seventy-nine, 79 **f** fifty-four, 54

Extra practice section: 1 a 12 **b** 18 **c** 15 **d** 21

C **1** b, c and f are coloured.
2 a, d and e are circled.

D **1 a** 8 triangles **b** 8 triangles **c** 21 squares **d** 9 rectangles

Unit 31 .... Page 78

A **1 a** 6 **b** 16 **c** 9 **d** 18 **e** 8 **f** 20

B **1 a** seventy-five, 75 **b** fifty-one, 51 **c** eighty-three, 83 **d** sixty-two, 62
e ninety-seven, 97 **f** forty-four, 44

Extra practice section: 1 $1.60

C **1 a** prism **b** cylinder **c** cone **d** cube (or prism) **e** sphere **f** pyramid

D **1 a** Friday **b** Sunday **c** Wednesday **d** Tuesday **e** Saturday
2 a 2nd **b** 25th **c** 7th **d** 29th

Unit 32 .... Page 80

A **1 a** 5 groups of 2 **b** 9 groups of 2 **c** 7 groups of 2 **d** 6 groups of 2 **e** 8 groups of 2

B **1 a** one half **b** one quarter **c** one half **d** one half **e** one quarter

C **1 a** square **b** circle **c** triangle **d** hexagon **e** rectangle

D **1 a** half past 10 **b** half past 2 **c** half past 9 **d** half past 5 **e** half past 11

Answers on page A5

C

1 Colour to match.

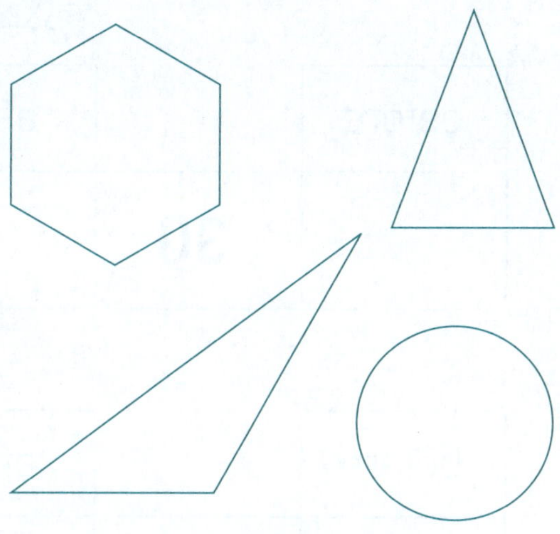

circle

triangle

hexagon

rectangle

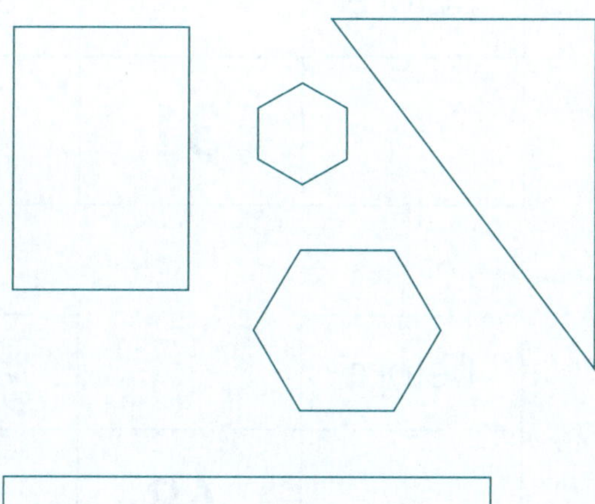

D

This is a hexagon.

1 Colour the smaller object.

a

b

c

d

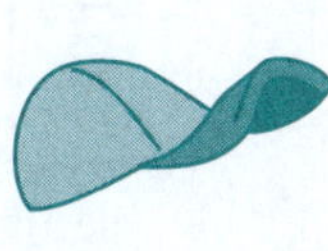

e

Unit 14

A

1 Complete.

a 

☐ – ☐ = ☐

b 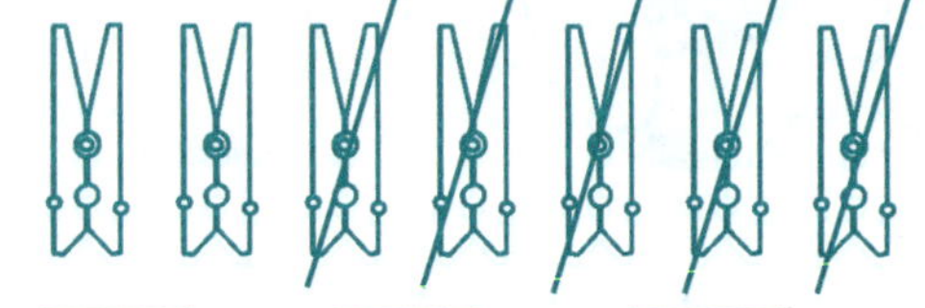

☐ – ☐ = ☐

c 

☐ – ☐ = ☐

d 

☐ – ☐ = ☐

e 

☐ – ☐ = ☐

B

1 Write the numbers.

a

before		after
	30	

b

before		after
	19	

c

before		after
	24	

d

before		after
	31	

e

before		after
	48	

Answers on page A5

C

1 Is the:

a chair in front of the table? ________

b fork next to the knife? ________

c cup near the jug? ________

d book behind the jug? ________

e knife behind the book? ________

f plate beside the knife? ________

D

1 Show the time.

a

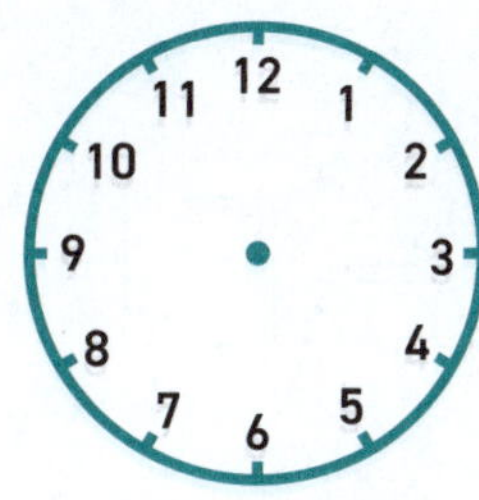

5 o'clock

b

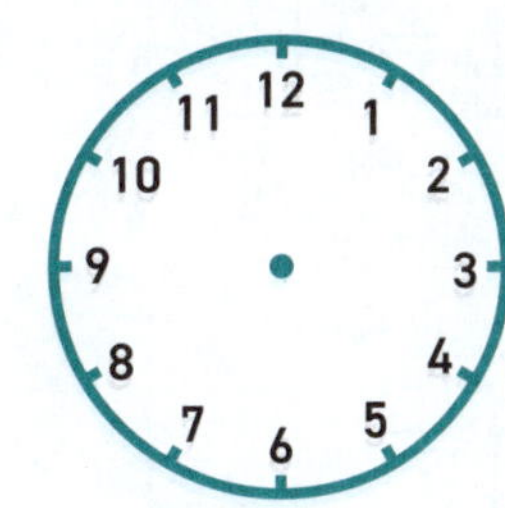

7 o'clock

c

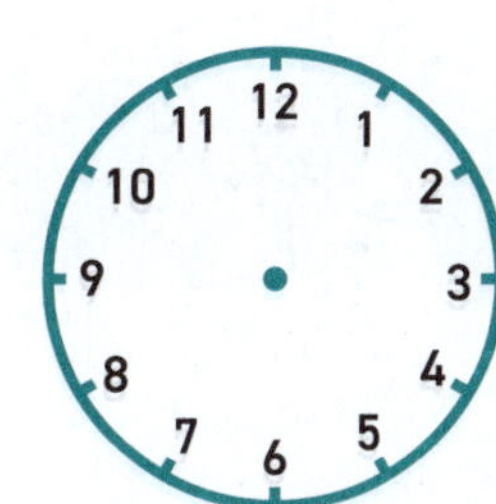

1 o'clock

d

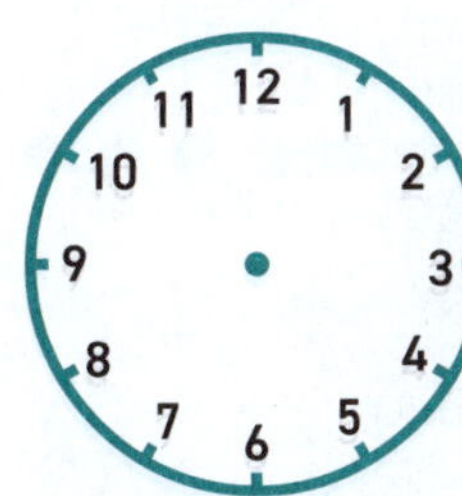

10 o'clock

e

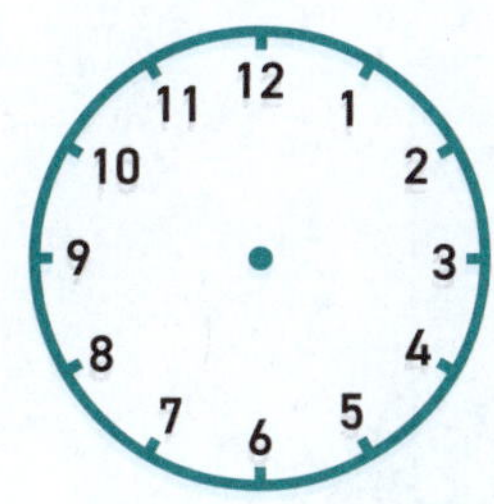

3 o'clock

Unit 15

A

1 Complete.

a 1 + 8 = ☐

b 5 + 4 = ☐

c 3 + 7 = ☐

d 6 + 3 = ☐

e 7 + 2 = ☐

f 9 + 1 = ☐

B

1 + 3 = 4

1 Write the numeral.

a twenty-five ______________

b sixteen ______________

c thirty-one ______________

d nineteen ______________

e thirty ______________

f forty-three ______________

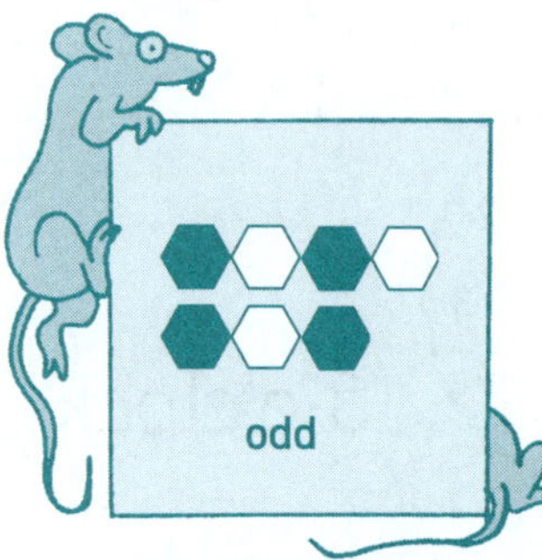

1 Colour the odd groups.

a ○○○○○ ○○○○

b ☐☐☐☐☐☐☐☐☐ ☐☐☐☐☐☐☐☐☐

c ⬡⬡⬡⬡⬡⬡⬡⬡ ⬡⬡⬡⬡⬡⬡⬡⬡⬡

Answers on page A5

C

1 How many:

a fish? ________

b frogs? ________

c dogs? ________

2 Are there more:

a fish than frogs? ________

b dogs than fish? ________

D

1 How many blocks?

a

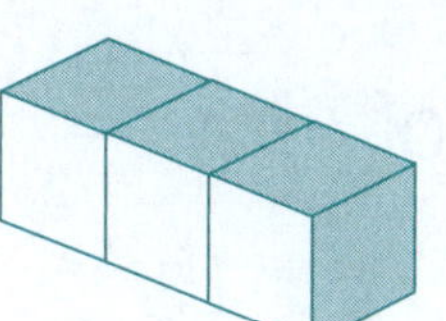

________ blocks

b

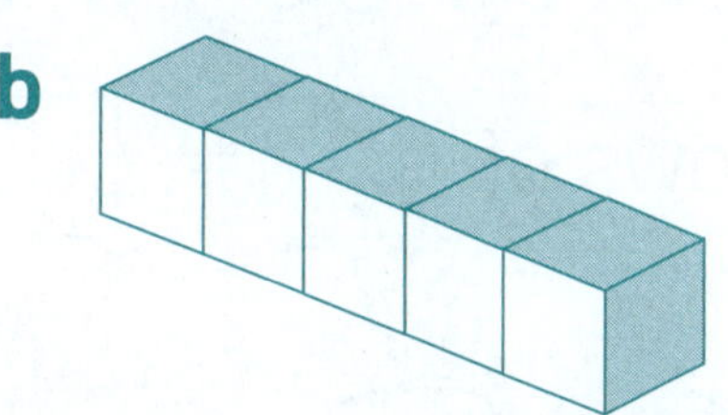

________ blocks

c

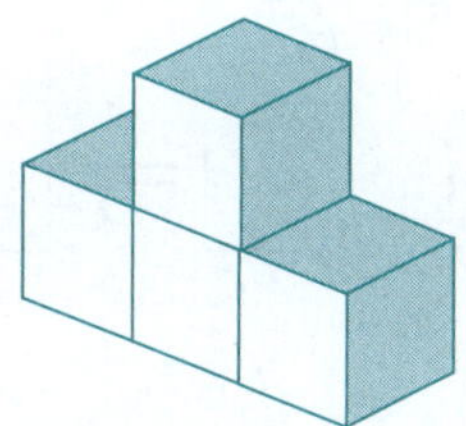

________ blocks

d

________ blocks

e

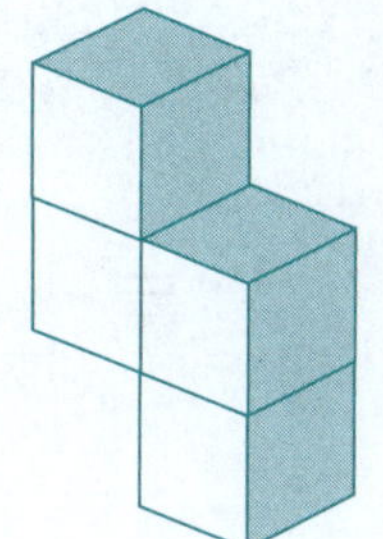

________ blocks

Unit 16

A

1 Complete.

a 

☐ rows of ☐ = ☐

b 

☐ rows of ☐ = ☐

c 

☐ rows of ☐ = ☐

d 

☐ rows of ☐ = ☐

e 

☐ rows of ☐ = ☐

B

1 Colour coins to show:

a 

30c

b 

35c

c 

65c

d 

50c

e 

70c

Answers on page A6

C

1 Colour the cubes.

a soap

b

c  Sudso

d peaches

e TOYS

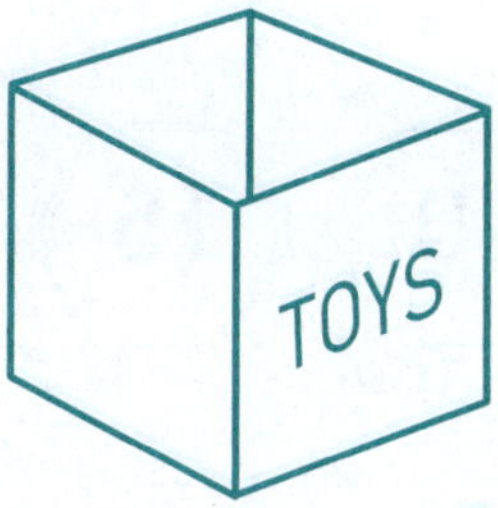

f

g

h

i

D

This is a cube.

1 Record the length.

a 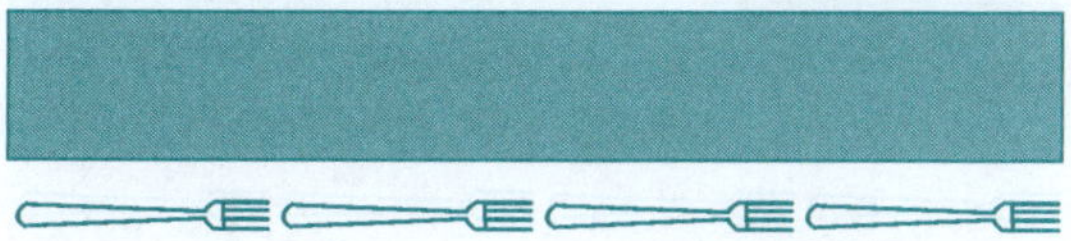 ______ forks

b ______ pencils

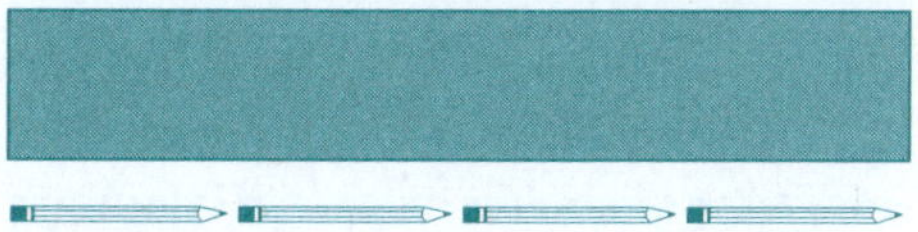

c ______ sticks

d ______ stars

e ______ blocks

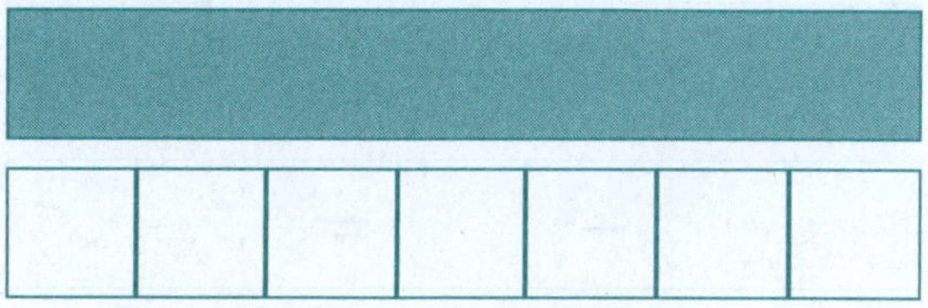

Unit 17

A

1 Complete.

a

___ – ___ = ___

b

___ – ___ = ___

c

___ – ___ = ___

d

___ – ___ = ___

e

___ – ___ = ___

B

1 Complete each number pattern.

a

1, 3, 5, ___

b 10, 9, 8, ___

c 5, 10, 15, ___

d 14, 16, 18, ___

e 13, 12, 11, ___

Answers on page A6

C

1 Colour the octagons.

a

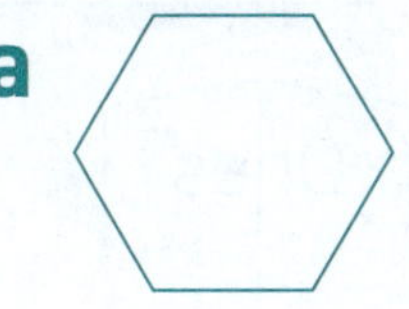

b

c

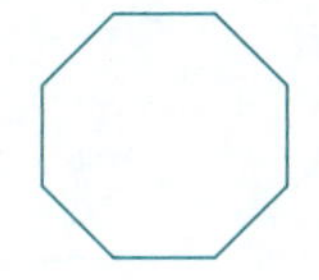

d

e

f

g

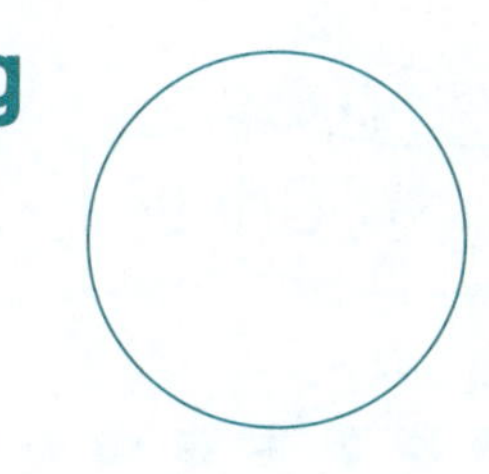

h

i

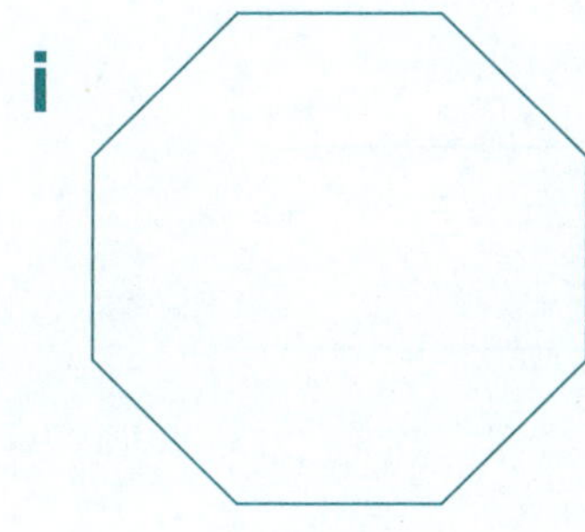

D

1 Write the missing days.

Sunday

Tuesday

Saturday

Unit 18

A

1 Complete.

a 5 − 3 = ☐

b 6 − 4 = ☐

c 7 − 2 = ☐

d 6 − 2 = ☐

e 8 − 1 = ☐

f 6 − 5 = ☐

B

1 Write the numeral.

a 3 Tens 2 Ones ______

b 1 Tens 5 Ones ______

c 2 Tens 8 Ones ______

d 5 Tens 3 Ones ______

e 3 Tens 7 Ones ______

f 6 Tens 1 Ones ______

1 Write the ordinal number for:

a fifth ______________

b ninth ______________

c seventh ______________

Answers on page A6

C

1 Colour the objects that look like each solid.

D

1 Write the month that follows.

a September

____

b May

____

c March

____

d November

____

e January

____

f July

____

A

1 Complete.

a

☐ rows of ☐ = ☐

b

☐ rows of ☐ = ☐

c

☐ rows of ☐ = ☐

d

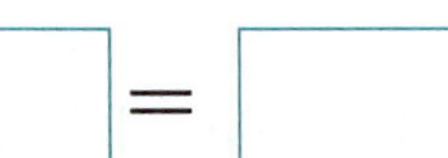

☐ rows of ☐ = ☐

e

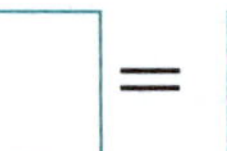

☐ rows of ☐ = ☐

B

1 Colour the fraction shown.

One half is coloured.

a one half

b

one quarter

c

one half

d

one half

e

one quarter

Answers on page A7

C

1 Colour the ovals.

a

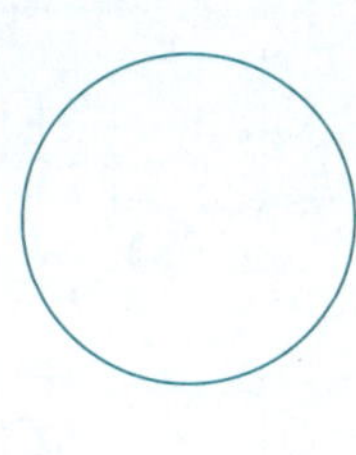

b

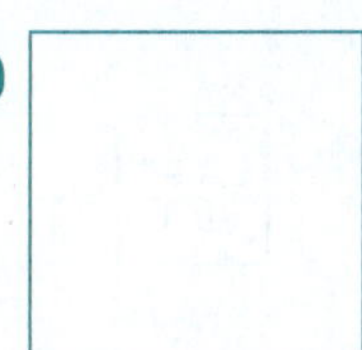

c

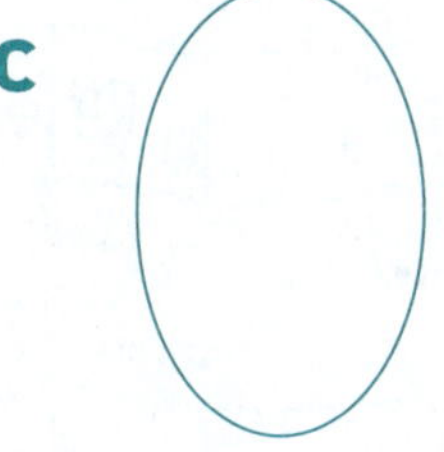

d

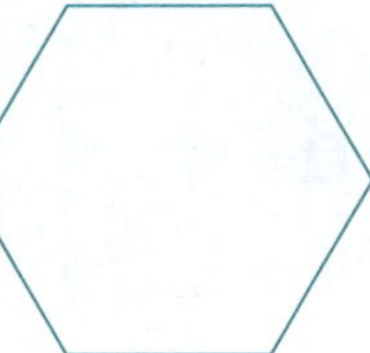

e

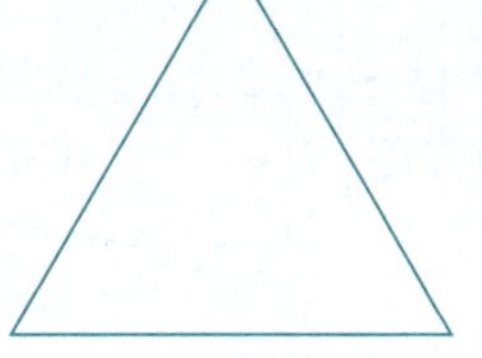

f

g

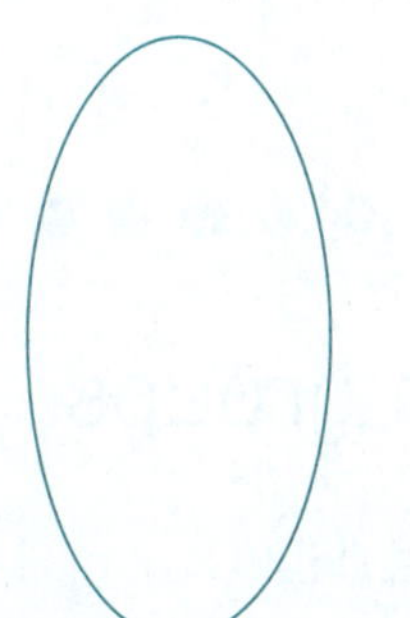

h

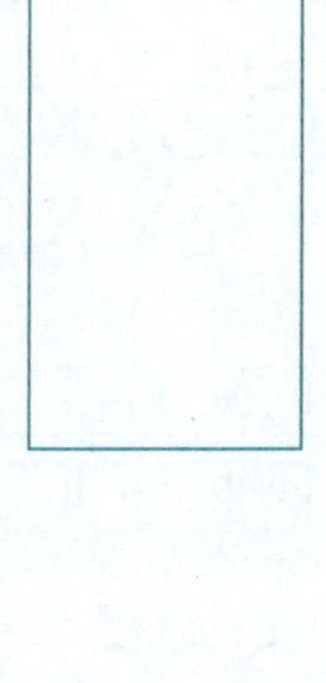

i

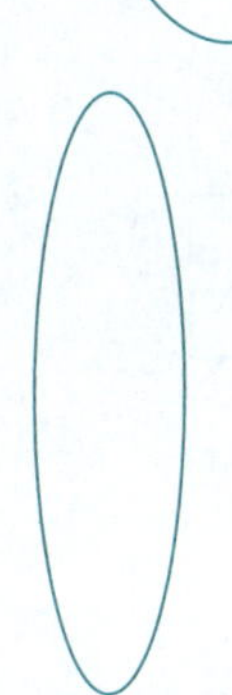

j

D

An oval is a flattened circle.

1 Show the time.

a

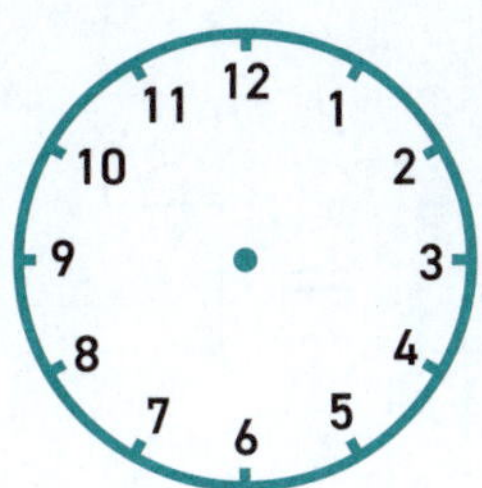

2 o'clock

b

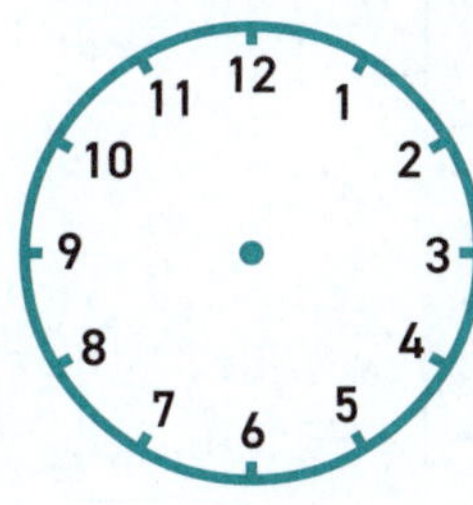

half past 4

c

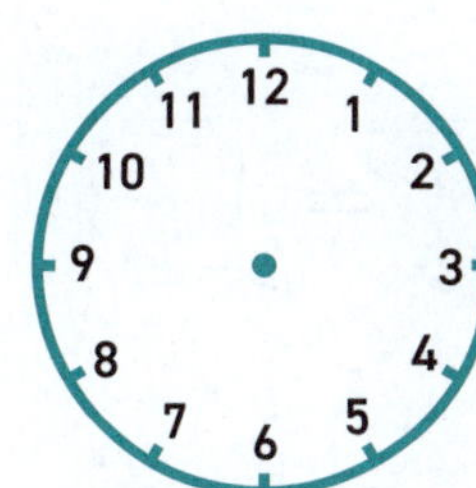

half past 8

d

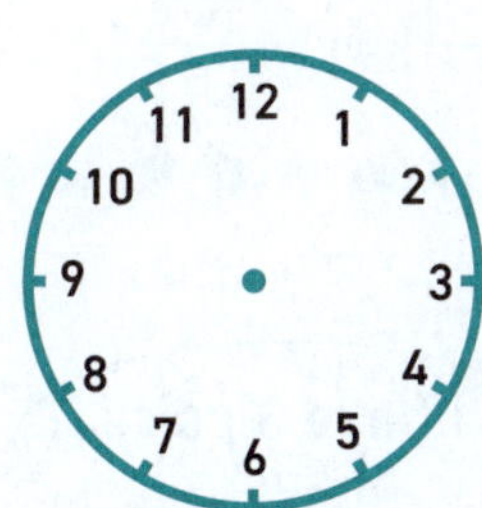

11 o'clock

e

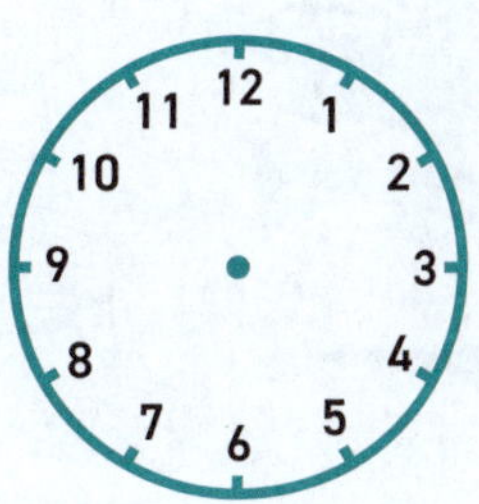

half past 9

Unit 20

A

1 Complete.

a $8 - 3 = \square$

b $7 - 4 = \square$

c $9 - 2 = \square$

d $6 - 3 = \square$

e $8 - 2 = \square$

f $7 - 3 = \square$

B

1 Complete the numeral expanders.

a 14 | ☐ Tens | ☐ Ones

b 39 | ☐ Tens | ☐ Ones

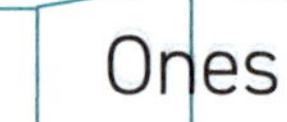

c 21 | ☐ Tens | ☐ Ones

d 56 | ☐ Tens | ☐ Ones

e 45 | ☐ Tens | ☐ Ones

f 73 | ☐ Tens | ☐ Ones

1 Colour the even groups.

a

b

c

Answers on page A7

C

1 Colour to match.

square

oval

octagon

hexagon

D

The area is 3 squares.

1 Record the area.

a

_____ rectangles

b

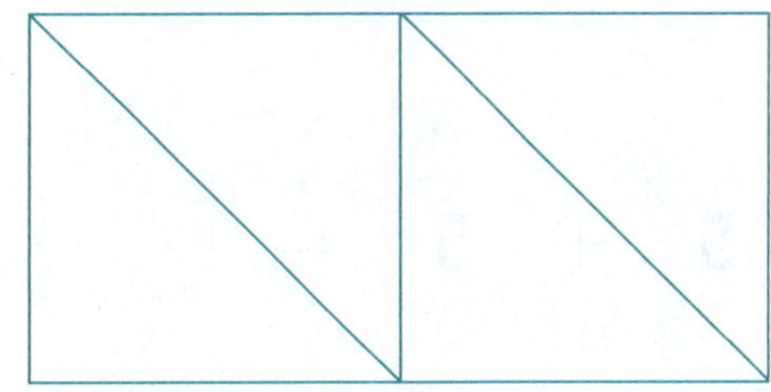

_____ triangles

c

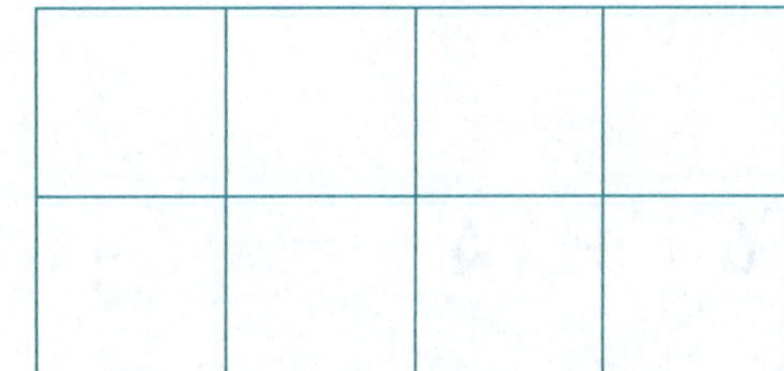

_____ squares

d

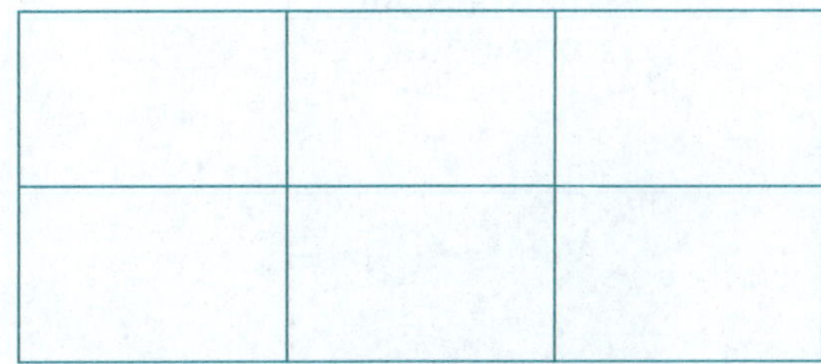

_____ rectangles

Unit 21

A

1 Complete.

a $1 + 7 = \square$

b $7 + 3 = \square$

c $4 + 6 = \square$

d $5 + 5 = \square$

e $8 + 2 = \square$

f $6 + 4 = \square$

B

The number line shows 5 + 5 = 10.

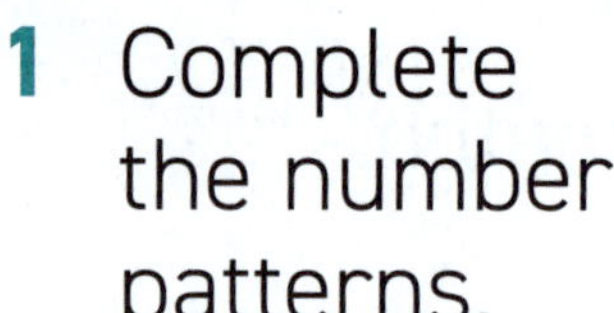

1 Complete the number patterns.

a

10, 12, 14, ☐

b

11, 13, 15, ☐

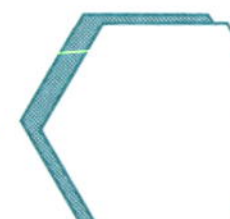

c

20, 19, 18, ☐

d
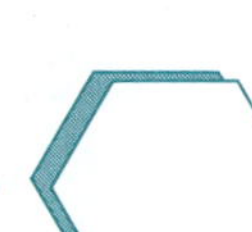

10, 15, 20, ☐

e

10, 8, 6, ☐

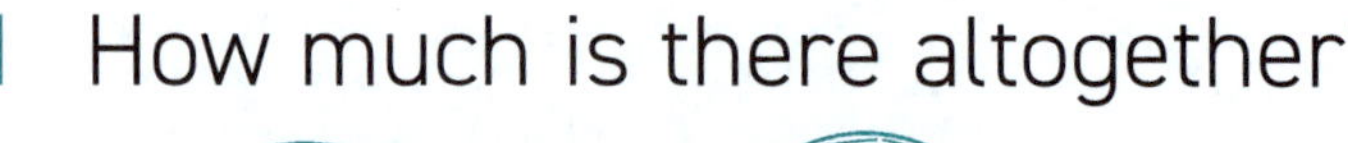

1 How much is there altogether?

☐ cents

Answers on page A8

C

1 How many sides has each shape?

2 Colour the quadrilaterals.

a
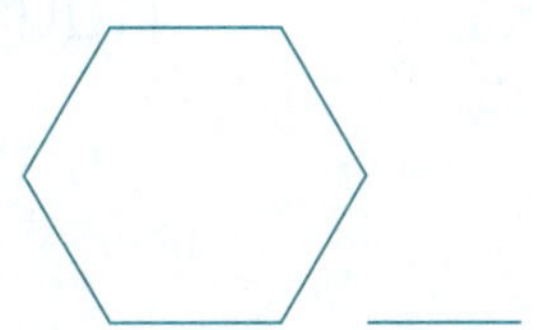
____

b
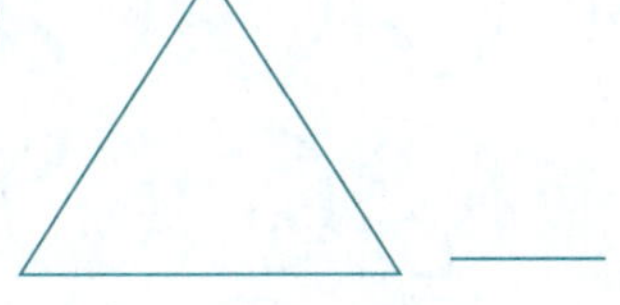
____

c

____

d

____

e

____

f
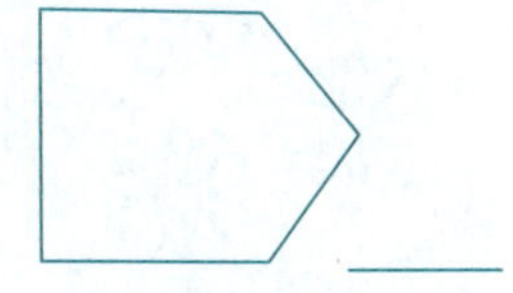
____

g
h
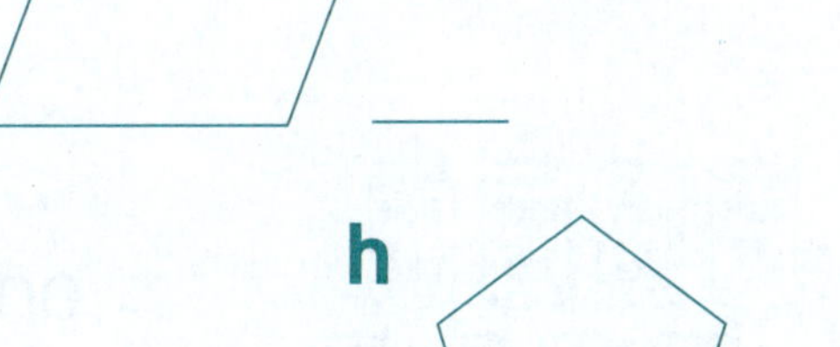
____ ____

D

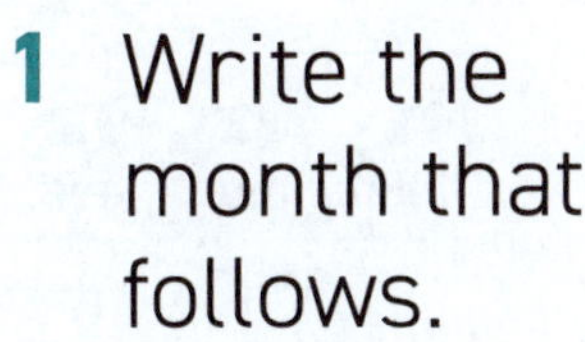

1 Write the month that follows.

a February

____

b December

____

c April

____

d August

____

e June

____

f October

____

Unit 22

2 groups of 4 = 8

A

1 Complete.

a

☐ groups of ☐ = ☐

b

☐ groups of ☐ = ☐

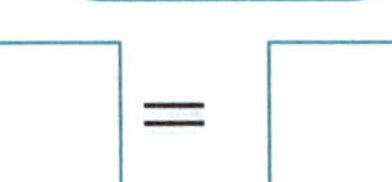

c

☐ groups of ☐ = ☐

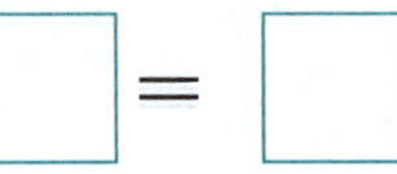

d

☐ groups of ☐ = ☐

e

☐ groups of ☐ = ☐

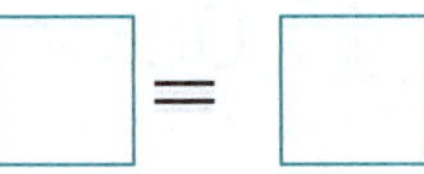

B

1 Colour the fraction shown.

a

one half

b

one quarter

c

one half

d

one half

e

one quarter

Answers on page A8

C

1 How many awards has:

a Paul? ________

b Chu? ________

c Sue? ________

d Rory? ________

2 Does Sue have:

a more than Chu? ________

b more than Rory? ________

D

1 How many blocks?

a

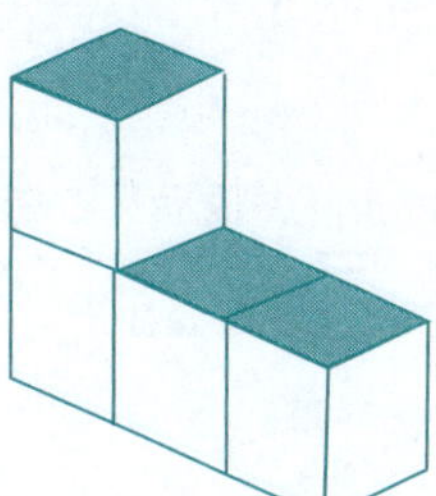

______ blocks

b

______ blocks

c

______ blocks

d

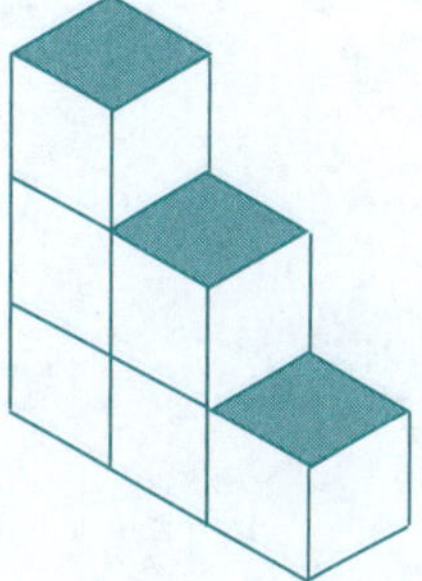

______ blocks

Unit 23

A

1 Complete.

a 7 − 5 = ☐

b 8 − 4 = ☐

c 9 − 3 = ☐

d 7 − 6 = ☐

e 5 − 5 = ☐

f 8 − 6 = ☐

B

The numeral for twenty-si is 26.

1 Write the numeral for:

a twenty-four ________

b fifty-nine ________

c thirty-two ________

d eighty-seven ________

e sixty-eight ________

f forty-six ________

1    

a How many plates are there? ________

b How many cakes are there? ________

c How many more plates than cakes are there? ________

Answers on page A8–A9

C

1 Complete each pattern.

a
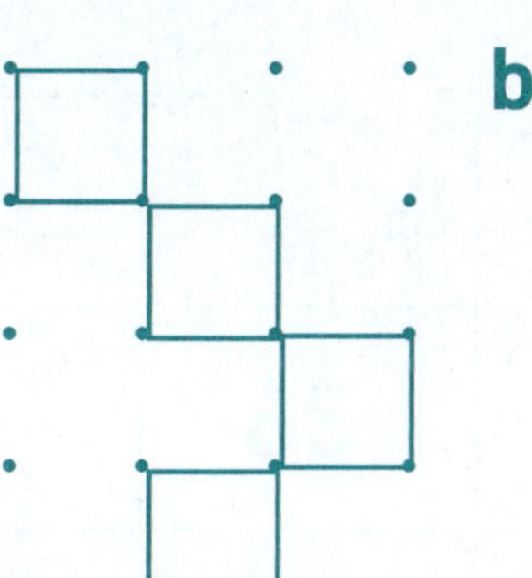

b
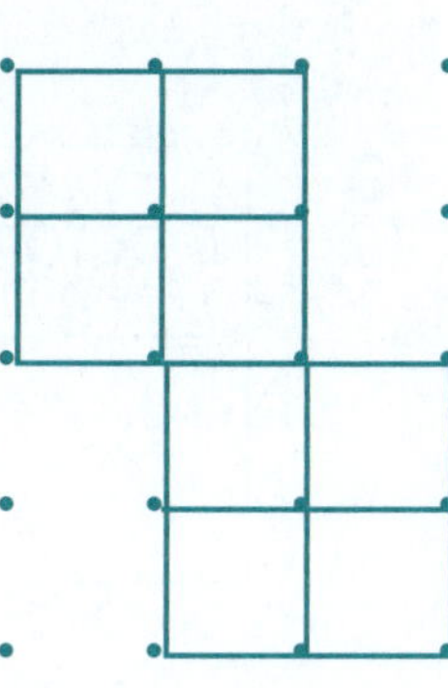

2 How many squares in pattern **a**? ___________

How many squares in pattern **b**? ___________

D

1 Write the time.

a

_______ o'clock

b

_______ o'clock

c

_______ o'clock

d

_______ o'clock

e

_______ o'clock

A

1 Is this a fair share?

a

______

b

______

c

______

d

______

e

______

B

1 Write the numbers.

a

before		after
	36	

b

before		after
	28	

c

before		after
	53	

d

before		after
	67	

e

before		after
	39	

Answers on page A9

C

1 Draw a fish inside the box.

2 Draw a flower behind the dog.

3 Draw a frog on the leaf.

4 Draw a boy beside the pigs.

5 Draw bubbles on the left of the fish.

D

I can measure area in squares.

1 Record the area.

a

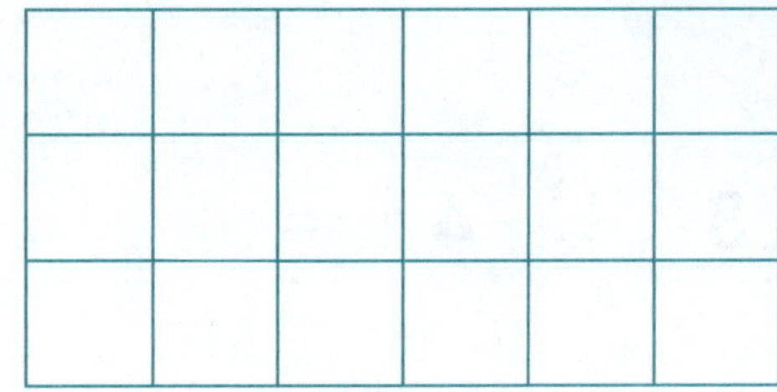

_______ squares

b

_______ rectangles

c

_______ triangles

d

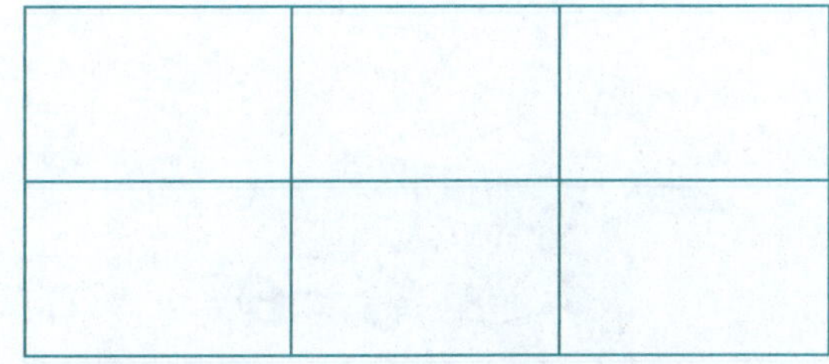

_______ rectangles

Unit 25

A

1 Complete.

a 9 + 2 = ☐

b 8 + 4 = ☐

c 7 + 4 = ☐

d 8 + 3 = ☐

e 9 + 3 = ☐

f 6 + 5 = ☐

B

1 Write the ordinal number for:

a eleventh ______

b fifteenth ______

c twelfth ______

d seventeenth ______

e sixteenth ______

f twentieth ______

1       

a How many saucers are there? ______

b How many cups are there? ______

c How many more saucers than cups are there? ______

Answers on page A9

C

1 Colour to match.

3 sides

4 sides

6 sides

D

That's the same as 2 o'clock.

2:00

1 Write the time.

a

________ o'clock

b

9:00

________ o'clock

c

________ o'clock

d

________ thirty

e

________ thirty

Unit 26

A

1 Complete.

a

☐ groups of ☐ = ☐

b

☐ groups of ☐ = ☐

c

☐ groups of ☐ = ☐

d

☐ groups of ☐ = ☐

e

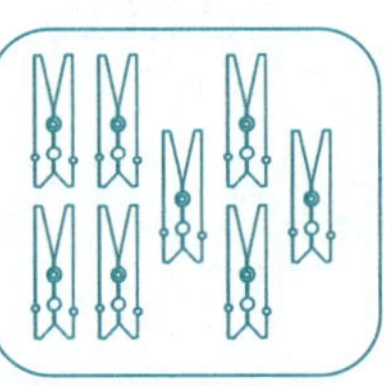
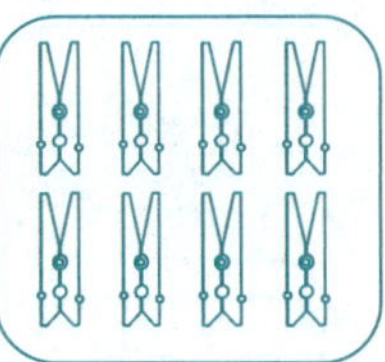

☐ groups of ☐ = ☐

B

1 Write the numeral for:

a

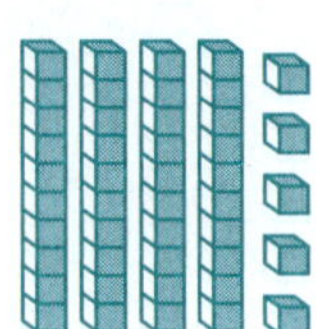

☐

b

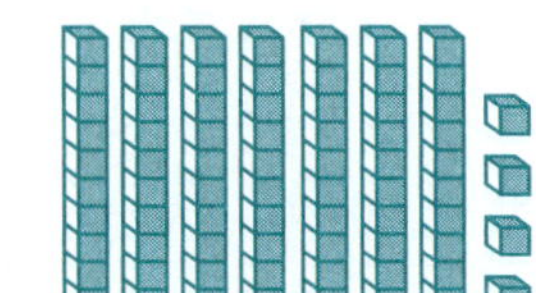

☐

c

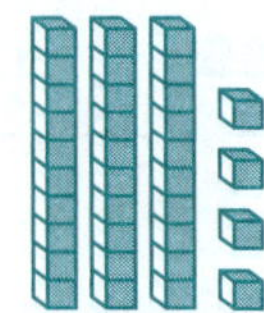

☐

d

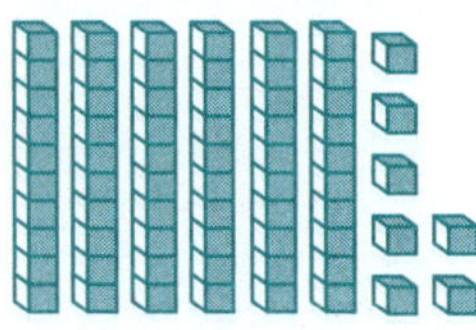

☐

e

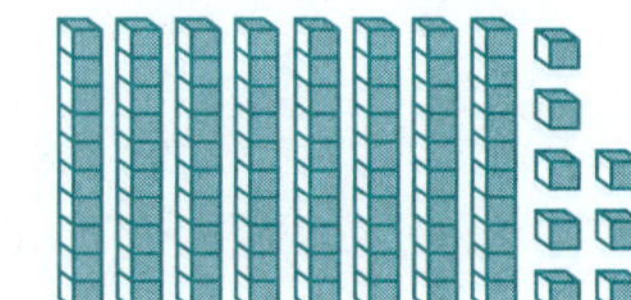

☐

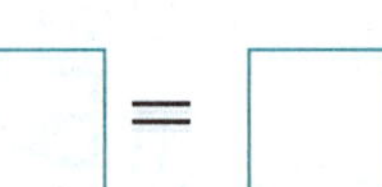

30 days has September, April ...

C

1 How many sides has each shape?

2 Colour the quadrilaterals.

a

____

b

____

c

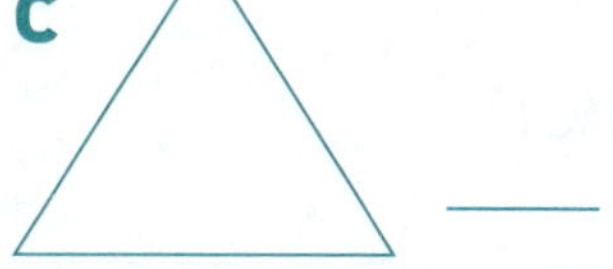

____

d

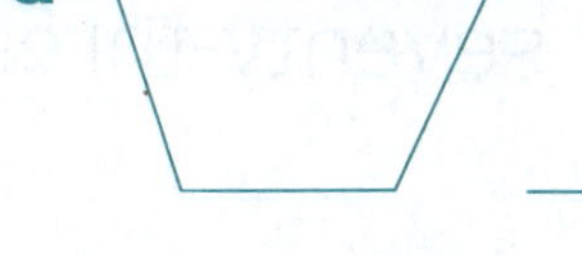

____

e

____

f ____

g ____

h

____

D

1 Colour the months that have exactly 30 days.

a January

b February

c March

d April

e May

f June

Unit 27

A

1 Complete.

a 9 − 4 = ☐

b 7 − 7 = ☐

c 3 − 3 = ☐

d 8 − 7 = ☐

e 9 − 6 = ☐

f 4 − 4 = ☐

B

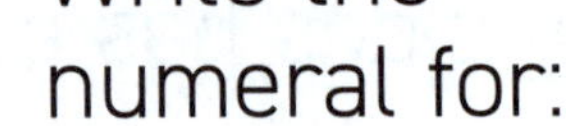
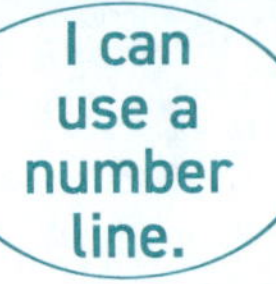

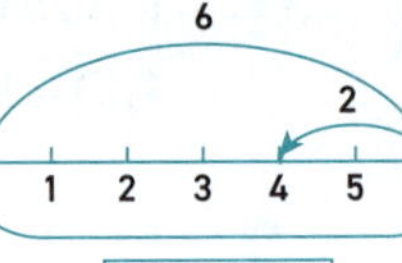

1 Write the numeral for:

a forty-five ______

b sixty-three ______

c fifty-eight ______

d twenty-seven ______

e seventy-three ______

f eighty-one ______

1        

a How many more fish than worms are there? ______

b 7 − 4 = ☐

Answers on page A10

C

1 Complete.

	Shape	Number of sides
a		
b		
c		
d 		
e	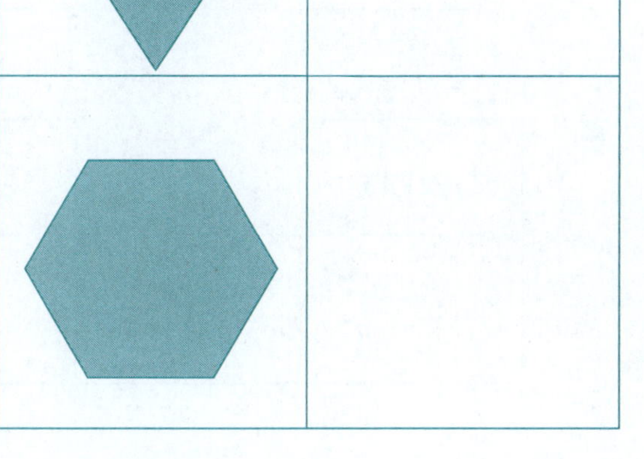	

D

Triangles have 3 corners.

1 Colour the months that have 31 days.

a July

b August

c September

d October

e November

f December

A

1 Is this a fair share?

a

______

b

______

c

______

d

______

e

______

B

Fair shares are equal.

1 Write the numbers.

a

before		after
	43	

b

before		after
	69	

c

before		after
	47	

d

before		after
	76	

e

before		after
	55	

Answers on page A10

C

1 Colour the curved surfaces.

2 Tick the faces.

a

b

c

d

e

f

D

A door is about a metre wide.

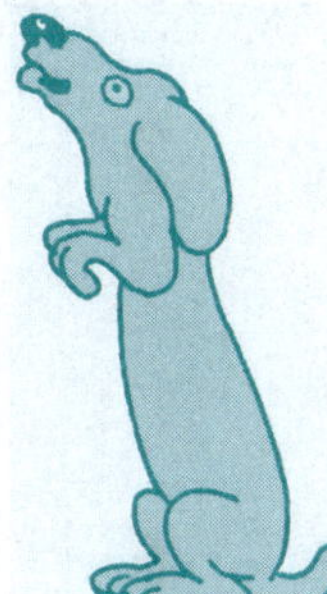

1 Colour objects that would be shorter than 1 metre.

a

b

c

d

e

f

g

Unit 29

A

1 Complete.

a 3 × 2 = ☐

b 6 × 2 = ☐

c 4 × 2 = ☐

d 2 × 2 = ☐

e 5 × 2 = ☐

f 7 × 2 = ☐

B

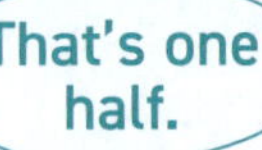

1 Colour the fraction shown: ●●○○

a one half

b one quarter

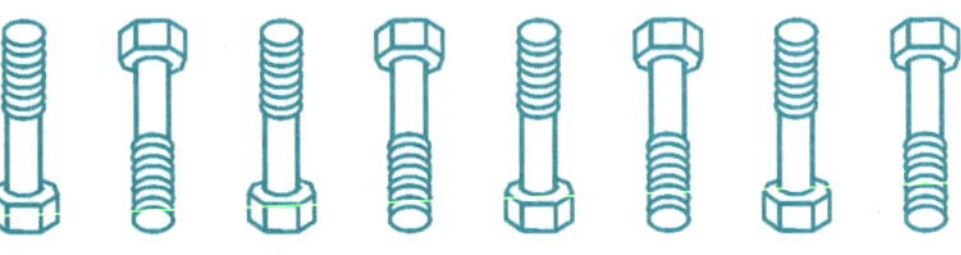

c one half

d one half

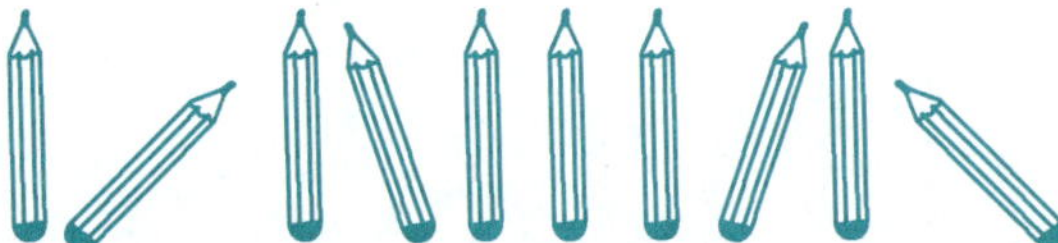

e one quarter

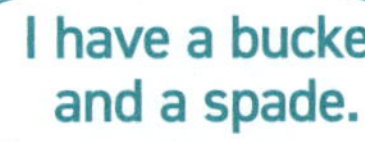

1

a How many more buckets than spades are there? ______

b 9 − 5 = ☐

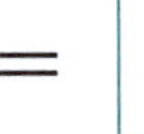

Answers on page A11

C

1 Name each solid shape.

a

____

b

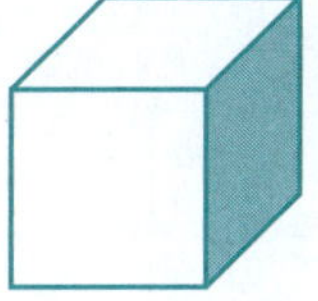

____

c

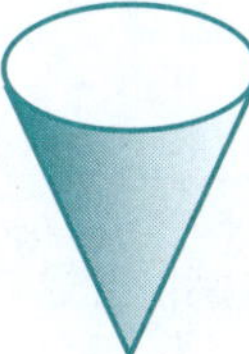

____

d

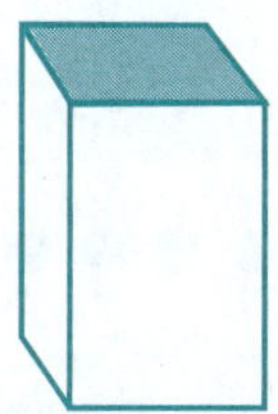

____

e

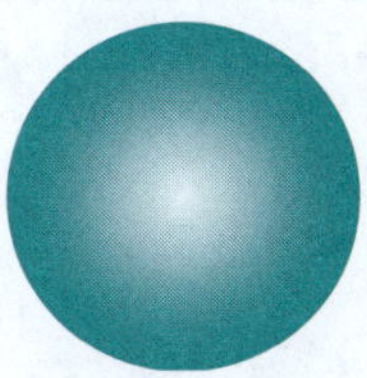

____

D

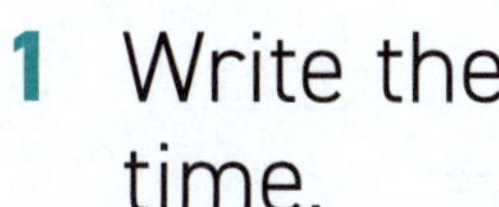

The big hand is on 6 for half past.

1 Write the time.

a

half past ________

b

half past ________

c

quarter past ________

d

half past ________

e

quarter past ________

Unit 30

A

1 Complete.

a 9 − 5 = ☐

b 6 − 6 = ☐

c 8 − 5 = ☐

d 9 − 7 = ☐

e 8 − 8 = ☐

f 9 − 8 = ☐

B

They match.

1 Colour to match.

twenty-two

	Word	Number
a	forty-eight	36
b	sixty-five	48
c	ninety-three	54
d	thirty-six	65
e	seventy-nine	93
f	fifty-four	79

1

★	★	★	★	★	★	★	★	★	★
★	★	★	★	★	★	★	★	★	★
★	★	★	★	★	★	★	★	★	★

a $3 \times 4 =$ ☐ b $3 \times 6 =$ ☐

c $3 \times 5 =$ ☐ d $3 \times 7 =$ ☐

Answers on page A11

C

1 Colour the pyramids.

2 Circle the prisms.

a

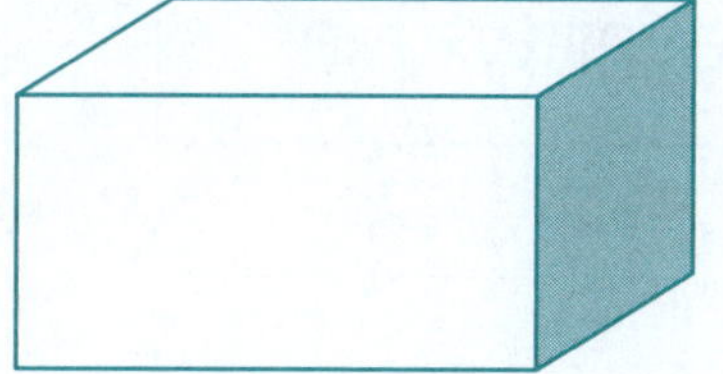

b

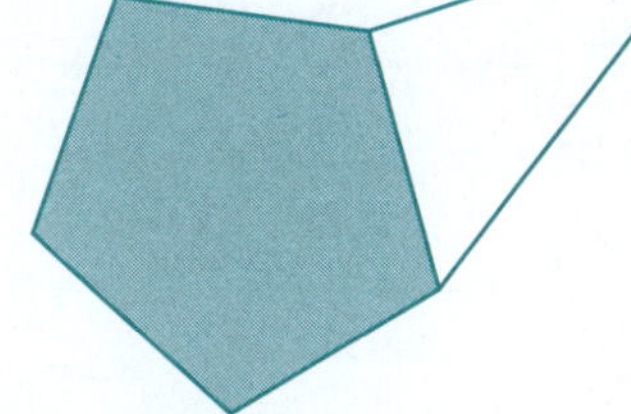

c

d

e

f

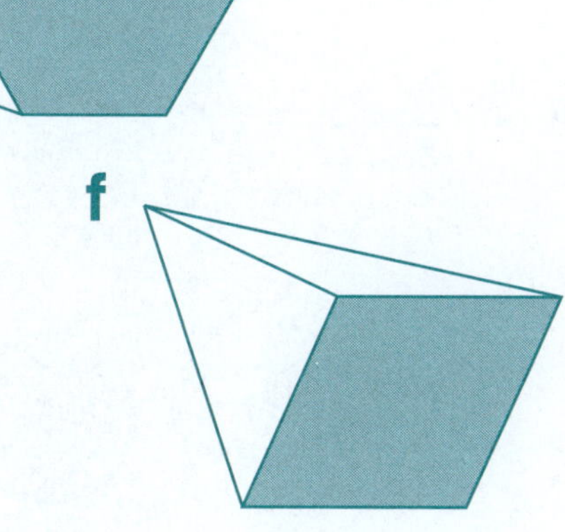

D

1 Record the area.

Pyramids have some faces that are triangles.

a

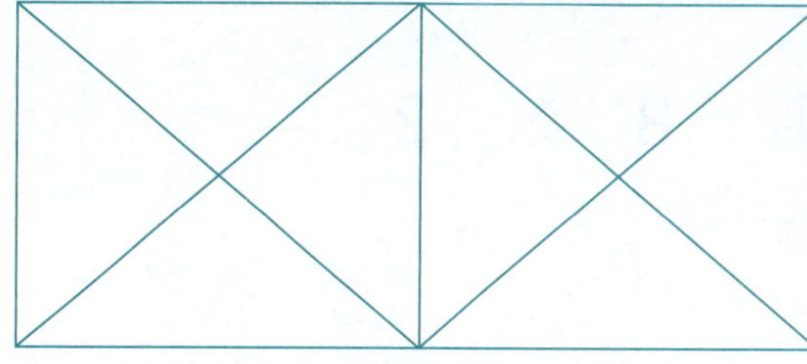

______ triangles

b

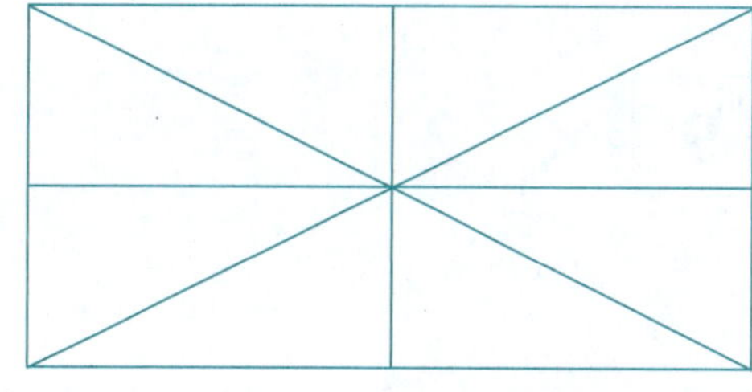

______ triangles

c

______ squares

d

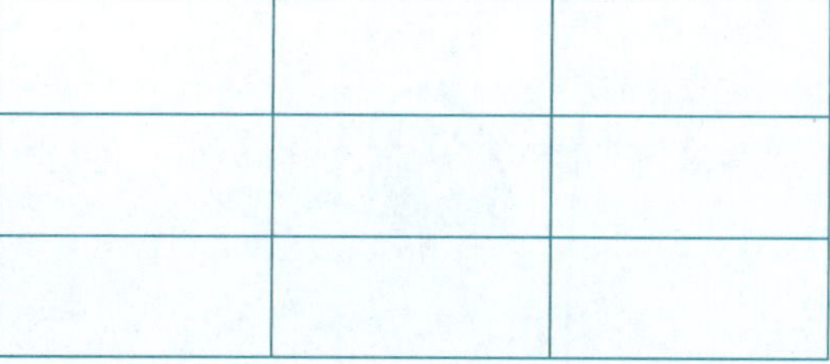

______ rectangles

Unit 31

A

1 Complete.

a $2 \times 3 = \square$

b $8 \times 2 = \square$

c $3 \times 3 = \square$

d $9 \times 2 = \square$

e $2 \times 4 = \square$

f $10 \times 2 = \square$

B

Fifty-six means the same as 56.

1 Colour to match.

a	seventy-five	83
b	fifty-one	75
c	eighty-three	97
d	sixty-two	51
e	ninety-seven	62
f	forty-four	44

1 How much is there altogether?

$ ________

Answers on page A12

C

1 Name each solid shape.

a ______________

b ______________

c  ______________

d ______________

e 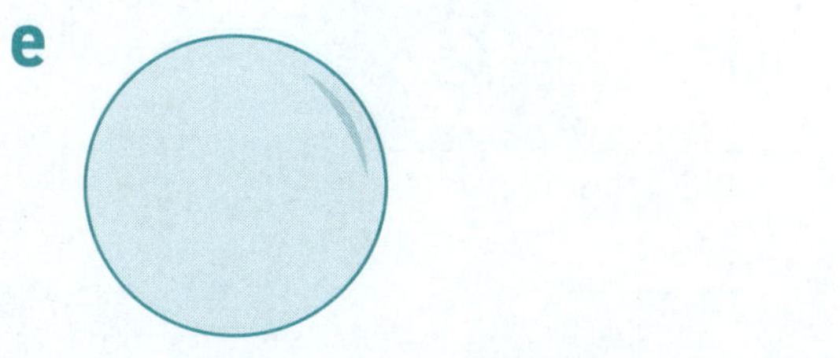 ______________

f  ______________

D

NOVEMBER

Sun	Mon	Tue	Wed	Thu	Fri	Sat
					1	2
3	4	5	6	7	8	9
10	11	12	13	14	15	16
17	18	19	20	21	22	23
24	25	26	27	28	29	30

1 What day of the week is:

a 1st? ______________

b 3rd? ______________

c 6th? ______________

d 19th? ______________

e 30th? ______________

2 Write the date for the:

a first Saturday ______________

b last Monday ______________

c first Thursday ______________

d last Friday ______________

Unit 32

A

1 Circle groups of 2.

a

_____ groups

b

_____ groups

c

_____ groups

d

_____ groups

e

_____ groups

B

Two groups of 2

1 Name the fraction coloured.

a

one _____

b

one _____

c

one _____

d

one _____

e

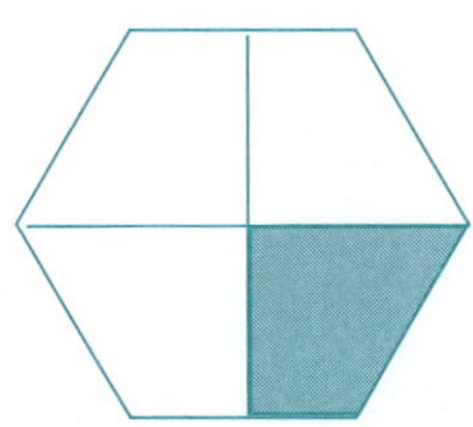

one _____

Answers on page A12

C

1 Name each shape.

a

____

b

____

c

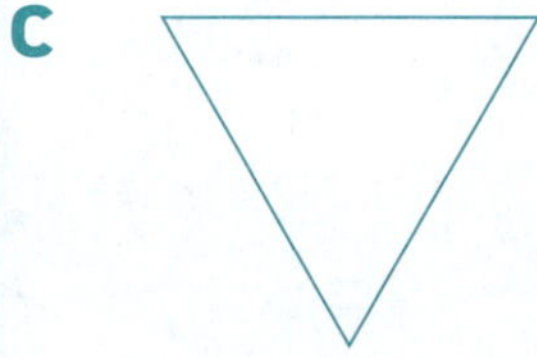

____

d

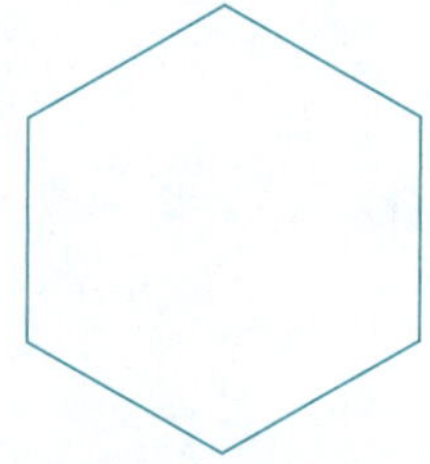

____

e

____

D

1 Write the time.

a

half past ________

b

half past ________

c

half past ________

d

half past ________

e

half past ________

Index to Help section

Notes